U.S.NRC
United States Nuclear Regulatory Commission
Protecting People and the Environment

I0482631

2012-2013
Information Digest

Nuclear Materials

Radioactive Waste

NRC

Nuclear Reactors

Nuclear Security

AVG. MEETING
4
A DAY

NUREG-1350, Volume 24
Manuscript Completed: July 2012
Date Published: August 2012
U.S. Nuclear Regulatory Commission
Office of Public Affairs
Washington, DC 20555-0001
www.nrc.gov

Cover and section spread art reflects infographics found within the Digest.

Cover Photos Front: (from left to right)
1. *Small ceramic fuel pellets*
2. *Inspectors examining blueprint*
3. *A Leskel Gamma Knife® headframe uses radiation beams to treat people with brain cancer. (Photo courtesy: Elekta)*
4. *Brunswick nuclear power plant (Photo courtesy: Progress Energy)*

Cover Photos Back: (from left to right)
1. *Armed security guard*
2. *Control room at a nuclear facility*
3. *Reactor vessel assembly*
4. *Positioning of a dry cask*

Mission

The mission of the U.S. Nuclear Regulatory Commission (NRC) is to license and regulate the Nation's civilian use of byproduct, source, and special nuclear materials to ensure adequate protection of public health and safety, to promote the common defense and security, and to protect the environment.

Commission

Chairman Allison M. Macfarlane
 Term Ends June 30, 2013
Commissioner Kristine L. Svinicki
 Term Ends June 30, 2017
Commissioner George Apostolakis
 Term Ends June 30, 2014
Commissioner William D. Magwood, IV
 Term Ends June 30, 2015
Commissioner William C. Ostendorff
 Term Ends June 30, 2016

Locations

Headquarters:
U.S. Nuclear Regulatory Commission
Rockville, MD, 301-415-7000, 1-800-368-5642
One White Flint North: 11555 Rockville Pike
Two White Flint North: 11545 Rockville Pike
Three White Flint North: 1106 Landsdown St.
 (anticipated occupation 11/2012)

Headquarters Operations Center:
Rockville, MD, 301-816-5100
The NRC maintains a staffed, 24-hour, Operations Center that is used to coordinate incident response concerns during an event with State, Local, and Federal agencies.

Regional Offices:
Region I
King of Prussia, PA
610-337-5000

Region II
Atlanta, GA
404-997-4000

Region III
Lisle, IL
630-829-9500

Region IV
Arlington, TX
817-860-8100

Training and Professional Development:
Technical Training Center, Chattanooga, TN
423-855-6500
Professional Development Center, Bethesda, MD
301-492-2000

Resident Sites:
At least two NRC resident inspectors, who report to the appropriate regional office, are located at each nuclear power plant site.

NRC Budget

- Total authority: $1,038 million
- Total staff: 3,953
- Budget amount expected to be recovered by annual fees to licensees: $909.5 million
- NRC research program support: $49.8 million

NRC Regulatory Activities

- Regulation and guidance—rulemaking
- Policymaking
- Licensing, decommissioning, and certification
- Research
- Oversight and enforcement
- Emergency preparedness and response
- Support of Commission decisions

NRC Governing Legislation

The NRC was established by the Energy Reorganization Act of 1974. A summary of laws that govern the agency's operations is provided below. NRC's regulations are found in Title 10 of the *Code of Federal Regulations*. The text of other laws may be found in NUREG-0980, "Nuclear Regulatory Legislation."

Fundamental Laws Governing Civilian Uses of Radioactive Materials

Nuclear Materials and Facilities
- Atomic Energy Act of 1954, as amended
- Energy Reorganization Act of 1974

Radioactive Waste
- Nuclear Waste Policy Act of 1982, as amended
- Low-Level Radioactive Waste Policy Amendments Act of 1985
- Uranium Mill Tailings Radiation Control Act of 1978

Nonproliferation
- Nuclear Non-Proliferation Act of 1978

Fundamental Laws Governing the Processes of Regulatory Agencies

- Administrative Procedure Act (5 U.S.C. Chapters 5 through 8)
- National Environmental Policy Act

Treaties and Agreements

- Nuclear Non-Proliferation Treaty
- International Atomic Energy Agency and U.S. Safeguards Agreement
- Convention on the Physical Protection of Nuclear Material
- Convention on Early Notification of a Nuclear Accident
- Convention on Assistance in Case of a Nuclear Accident and Radiological Emergency
- Convention on Nuclear Safety
- Joint Convention on the Safety of Spent Fuel Management and the Safety of Radioactive Waste Management

U.S. Commercial Nuclear Power Reactors

- Generate about 20 percent of the Nation's electricity
- 31 States with operating reactors
- 104 nuclear power plants licensed to operate in the United States: 69 pressurized-water reactors 35 boiling-water reactors
- 4 reactor fuel vendors
- 26 parent companies
- 80 different designs
- 65 commercial reactor sites
- 17 power reactors undergoing decommissioning
- 6,820 total inspection hours at operating reactors in calendar year (CY) 2011
- Approximately 3,000 inspection documents concerning events reviewed

Reactor License Renewal

Commercial power reactor operating licenses are valid for 40 years and may be renewed for up to an additional 20 years.

- 31 units with original license
- 44 sites comprised of 73 units issued renewal licenses
- 9 sites with license renewal applications in review
- 11 sites with letters of intent to submit renewal license applications

New Reactor License Process

Early Site Permit (ESP)

- 4 ESPs issued and 2 applications in review

Combined License—Construction and Operating (COL)

- 4 COLs issued and 16 applications received and docketed for 24 units; of these, 10 applications are under active review

Reactor Design Certification (DC)

- 4 DCs issued and 3 applications in review

Nuclear Research and Test Reactors

42 licensed research reactors and test reactors

- 31 reactors operating in 21 States
- 11 reactors permanently shut down and in various stages of decommissioning (since 1958, a total of 83 licensed research and test reactors have been decommissioned)

Nuclear Security and Safeguards

- Once every 2 years, each nuclear power plant performs full-scale emergency preparedness exercises.
- Plants also conduct additional emergency drills between full-scale exercises. The NRC and FEMA evaluate emergency exercises and drills.

Nuclear Materials

- The NRC and the Agreement States have issued 21,800 licenses for medical, academic, industrial, and general uses of nuclear materials a year.
- The NRC oversees approximately 3,000 licenses.
- 37 Agreement States oversee approximately 18,900 licenses.

18 Uranium Recovery Sites Licensed by the NRC

- 7 in situ recovery sites
- 11 conventional mills (10 undergoing decommissioning)

15 Fuel Cycle Facilities

- 1 uranium hexafluoride production facility
- 6 uranium fuel fabrication facilities
- 1 gaseous diffusion uranium enrichment facility
- 3 gas centrifuge uranium enrichment facilities (1 operating with further construction and 1 under construction)
- 1 mixed-oxide fuel fabrication facility (under construction and review)
- 1 laser separation enrichment facility (under review)
- 1 uranium hexaflouride deconversion facility (under review)
- 180 NRC-licensed facilities authorized to possess plutonium and enriched uranium with inventory registered in the Nuclear Material Management and Safeguards System database

Radioactive Waste

Low-Level Radioactive Waste

- 10 regional compacts
- 4 active licensed disposal facilities
- 4 closed disposal facilities

High-Level Radioactive Waste Management

Disposal and Storage

There are no active high-level radioactive waste disposal facilities. In September 2011, the NRC completed an orderly closure of its Yucca Mountain, NV, activities.

Spent Nuclear Fuel Storage

- 65 licensed and/or operating independent spent fuel storage installations in 34 States
- 15 site-specific licenses
- 50 general licenses

Transportation—Principal Licensing and Inspection Activities

- The NRC examines transport-related safety during approximately 1,000 safety inspections of fuel, reactor, and materials licensees annually.
- The NRC reviews, evaluates, and certifies approximately 80 new, renewal, or amended container-design applications for the transport of nuclear materials annually.

Facts at a Glance

- The NRC reviews and evaluates approximately 150 license applications for the import and export of nuclear materials from the United States annually.
- The NRC inspects about 28 dry storage and transport package licensees annually.

Decommissioning

Approximately 150 materials licenses are terminated each year. The NRC's decommissioning program focuses on the termination of licenses that are not routine and that require complex activities.

- 29 nuclear power reactors permanently shut down
- 12 nuclear reactors completely decommissioned and licenses terminated
- 17 nuclear reactors in various stages of decommissioning (DECON, SAFSTOR, or ENTOMB)
- 11 research and test reactors
- 18 complex material sites
- 1 fuel cycle facility (partial decommissioning)
- 11 uranium recovery facilities in safe storage under NRC jurisdiction

Public Meetings and Involvement

- The NRC conducts more than 1,000 public meetings annually.
- The NRC hosts both the Regulatory Information Conference and the Fuel Cycle Information Exchange annually, where thousands of participants from around the world discuss the latest technical issues.
- The Advisory Committee on Reactor Safeguards held 10 full committee meetings and approximately 70 subcommittee meetings in CY 2011.
- The Advisory Committee on Medical Uses of Isotopes typically holds public meetings twice a year.

News and Information

- NRC news releases are available through a free listserv subscription at www.nrc.gov/public-involve/listserver.html.
- The NRC uses social media as a communication tool to allow the public to stay connected through the NRC Blog, Twitter, Flickr, and YouTube.

NRC Regions

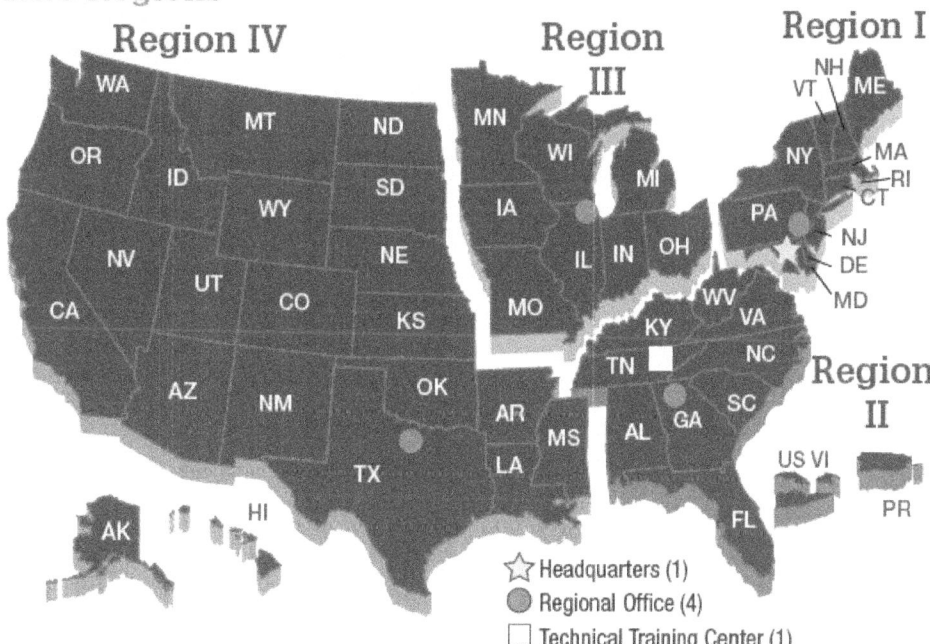

Region IV · Region III · Region I · Region II

☆ Headquarters (1)
⬤ Regional Office (4)
☐ Technical Training Center (1)

Nuclear Power Plants
- Each regional office oversees the plants in its region, except the Grand Gulf plant in Mississippi and the Callaway plant in Missouri, which Region IV oversees.

Materials Licensees
- Region I oversees licensees and Federal facilities located geographically in Region I and Region II.
- Region III oversees licensees and Federal facilities located geographically in Region III.
- Region IV oversees licensees and Federal facilities located geographically in Region IV.

Nuclear Fuel Processing Facilities
- Region II oversees all the fuel processing facilities in the region and those in Illinois, New Mexico, and Washington.
- In addition, Region II handles all construction inspectors' activities for new nuclear power plants and fuel cycle facilities in all regions.

Contact Us

Mailing Address

U.S. Nuclear Regulatory Commission
Washington, DC 20555-0001
1-800-368-5642, 301-415-7000, TTD: 301-415-5575

Delivery Address

11555 Rockville Pike, Rockville, MD 20852

Public Affairs

301-415-8200, Fax: 301-415-3716

Public Document Room

1-800-397-4209, Fax: 301-415-3548
TDD: 1-800-635-4512

Employment

Human Resources 301-415-7400

Office of the General Counsel Intern Program or Honor Law Graduate Programs 301-415-1515

Contracting Opportunities

Small Business Office 1-800-903-7227

Report a Concern

Emergency

Involving a nuclear facility or radioactive materials, including the following:

- any accident involving a nuclear reactor, nuclear fuel facility, or radioactive materials,
- lost or damaged radioactive materials,
- any threat, theft, smuggling, vandalism, or terrorist activity involving a nuclear facility or radioactive materials.

**Call the NRC's 24-Hour
Headquarters Operations Center:
301-816-5100**

We accept collect calls, and all calls to this number are recorded.

Non-Emergency

Including any concern involving a nuclear reactor, nuclear fuel facility, or radioactive materials.

You may send an e-mail to Allegations@nrc.gov. However, because e-mail transmission may not be completely secure, if you are concerned about protecting your identity, it is preferable that you contact us by telephone or in person. You may contact any NRC employee (including a resident inspector) or call:

**Call the NRC's Toll-Free Safety Hotline:
800-695-7403**

Calls to this number **are not recorded between the hours of 7:00 a.m. and 5:00 p.m. EST.** However, calls received outside these hours are answered by the Incident Response Operations Center on a recorded line.

Some materials and activities are regulated by Agreement States, and concerns should be directed by contacting the appropriate State Radiation Control Program.

NRC's Office of the Inspector General

The Office of the Inspector General (OIG) at the NRC established the Hotline program to provide the NRC employee, other government employee, licensee and utility employee, contractor employee, and the public with a confidential means of reporting incidences of suspicious activity to OIG concerning fraud, waste, abuse, and employee or management misconduct. Mismanagement of agency programs or danger to public health and safety may also be reported through the Hotline.

It is not OIG policy to attempt to identify people contacting the Hotline. People may contact OIG by telephone, through an online form, or by mail. There is no caller identification feature associated with the Hotline or any other telephone line in the Inspector General's office. No identifying information is captured when you submit an online form. You may provide your name, address, or telephone number, if you wish.

**Call the OIG Hotline:
1-800-233-3497, TDD: 1-800-270-2787
7:00 a.m.–4:00 p.m. (EST)
After hours, please leave a message.**

Stay Connected

http://public-blog.nrc-gateway.gov/

https://twitter.com/#!/nrcgov

http://www.youtube.com/user/NRCgov

http://www.flickr.com/photos/nrcgov

http://www.nrc.gov/public-involve/listserver.htm#rss

Abstract

The U.S. Nuclear Regulatory Commission (NRC) 2012–2013 Information Digest provides a summary of information about the NRC and the industries it regulates. It describes the agency's regulatory responsibilities and licensing activities, and provides general information on nuclear-related topics. It is updated annually. The Information Digest includes NRC and industry data in a quick-reference format for activities through 2011 or most recent current data available at manuscript completion. InfoGraphics have been incorporated to help provide a visual representation of information expanding the agency's efforts to be more transparent and broaden its public outreach. The Web Link Index provides Web addresses for more information on major topics. The Digest also includes a tear-out reference sheet the NRC Facts at a Glance. The NRC reviewed information from industry and international sources but did not perform an independent verification. This edition contains adjustments to preliminary figures from the previous year. All information is final unless otherwise noted. The NRC is the source for all photographs, graphics, and tables unless otherwise noted. The agency welcomes comments or suggestions on the Information Digest. Please contact the Office of Public Affairs, by mail at U.S. Nuclear Regulatory Commission, Washington, DC 20555-0001, by e-mail at OPA.Resource@nrc.gov or post comments on the NRC Blog at http://public-blog.nrc-gateway.gov/.

Contents

Appendices

Contents

Web Link Index

Index

Figures

NRC: an Independent Regulatory Agency

U.S. and Worldwide Nuclear Energy

Nuclear Reactors

Nuclear Materials

Tables

NRC: an Independent Regulatory Agency

1,040
Public Meetings in 2011

527
Category 1:
Observation

231
Category 2:
Feedback

130
Category 3:
Fully Engaged

151
Open Meetings
(ACRS, ASLBP, ACMUI
Commission Briefings)

1 Closed

AVG. MEETINGS
4
A DAY

53
WEBCAST
MEETINGS

THE ANNUAL
**REGULATORY
INFORMATION
CONFERENCE**

3,000
ATTENDEES FROM
30 DIFFERENT
COUNTRIES

= 20 people

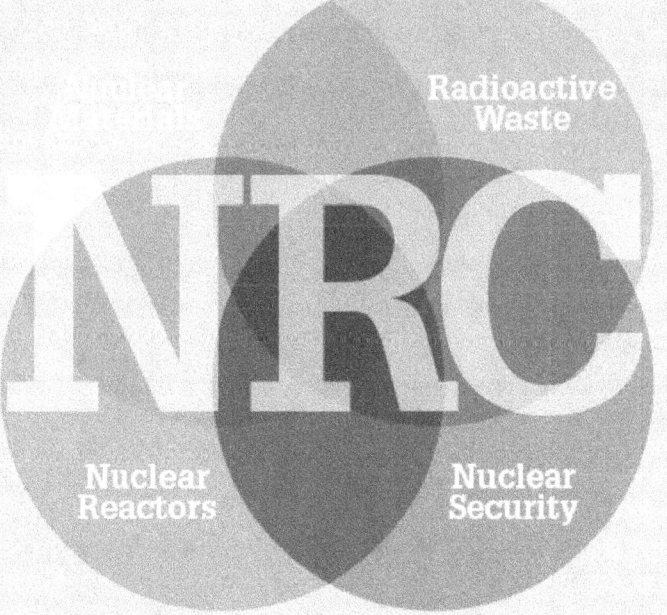

Region IV

WA
OR
MT
ND
ID
SD
WY
NE
NV
UT
CO
KS
CA
AZ
NM
OK
TX
AK
HI

Region III

MN
WI
MI
IA
IL
IN
OH
MO
AR
MS
LA

Region I

NH
VT
ME
NY
MA
RI
CT
PA
NJ
DE
MD

WV
VA
KY
TN
NC
SC

Region II

AL
GA
US VI
FL
PR

☆ Headquarters (1)
● Regional Office (4)
□ Technical Training Center (1)

NRC

Nuclear Materials

Radioactive Waste

Nuclear Reactors

Nuclear Security

Mission

The U.S. Nuclear Regulatory Commission (NRC) is an independent agency created by Congress. The mission of the NRC is to license and regulate the Nation's civilian use of byproduct, source, and special nuclear materials to ensure the adequate protection of public health and safety, promote the common defense and security, and protect the environment. The NRC's regulations are designed to protect both the public and workers against radiation hazards from industries that use radioactive materials. The NRC's scope of responsibility includes regulation of commercial nuclear power plants; research, test, and training reactors; nuclear fuel cycle facilities; medical, academic, and industrial uses of radioactive materials; and the transport, storage, and disposal of radioactive materials and wastes. In addition, the NRC licenses the import and export of radioactive materials and works to enhance nuclear safety and security throughout the world.

Values

The NRC adheres to the principles of good regulation—independence, openness, efficiency, clarity, and reliability. The agency puts these principles into practice with effective, realistic, and timely regulatory actions.

Strategic Goals

Safety: Ensure adequate protection of public health and safety and the environment.

Security: Ensure adequate protection in the secure use and management of radioactive materials.

Strategic Outcomes

- Prevent the occurrence of any nuclear reactor accidents.
- Prevent the occurrence of any inadvertent criticality events.
- Prevent the occurrence of any acute radiation exposures resulting in fatalities.
- Prevent the occurrence of any releases of radioactive materials that result in significant radiation exposures.
- Prevent the occurrence of any releases of radioactive materials that cause significant adverse environmental impacts.
- Prevent any instances where licensed radioactive materials are used domestically in a manner hostile to the United States.
- Prevent unauthorized public disclosures of classified or Safeguards Information through quality measures.

Figure 1. How We Regulate

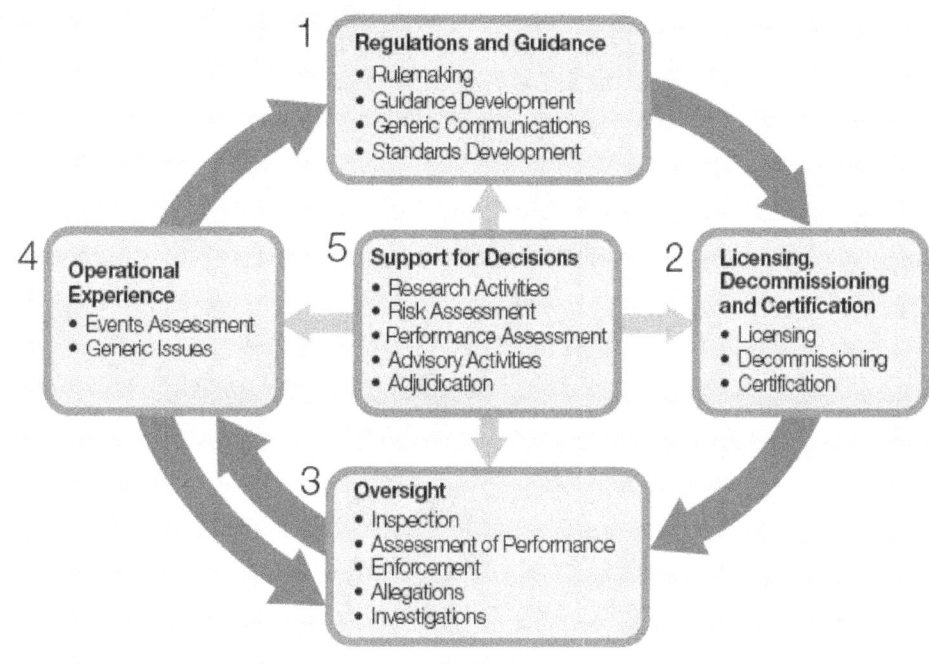

1
Regulations and Guidance
- Rulemaking
- Guidance Development
- Generic Communications
- Standards Development

4
Operational Experience
- Events Assessment
- Generic Issues

5
Support for Decisions
- Research Activities
- Risk Assessment
- Performance Assessment
- Advisory Activities
- Adjudication

2
Licensing, Decommissioning and Certification
- Licensing
- Decommissioning
- Certification

3
Oversight
- Inspection
- Assessment of Performance
- Enforcement
- Allegations
- Investigations

1. Developing regulations and guidance for applicants and licensees.
2. Licensing or certifying applicants to use nuclear materials, operate nuclear facilities, and decommission facilities.
3. Inspecting and assessing licensee operations and facilities to ensure that licensees comply with NRC requirements, investigating allegations of wrong-doing and taking appropriate followup or enforcement actions when necessary.
4. Evaluating operational experience of licensed facilities and activities.
5. Conducting research, holding hearings, and obtaining independent reviews to support regulatory decisions.

Statutory Authority

The Energy Reorganization Act of 1974 established the NRC to oversee the commercial nuclear industry. The agency took over regulation formerly carried out by the Atomic Energy Commission and began operations on January 18, 1975. As previously noted, the NRC regulates the civilian commercial, industrial, academic, and medical uses of nuclear materials. Effective regulation enables the Nation to use radioactive materials for beneficial civilian purposes while protecting the American people and their environment. The NRC's regulations are contained in Title 10, "Energy," of the *Code of Federal Regulations* (10 CFR). The following principal statutory authorities govern the NRC's work and can be found on the NRC Web site (see the Web Link Index):

- Atomic Energy Act of 1954, as amended (Pub. L. 83–703)

- Energy Reorganization Act of 1974, as amended (Pub. L. 93–438)

- Uranium Mill Tailings Radiation Control Act of 1978, as amended (Pub. L. 95–604)

- Nuclear Non-Proliferation Act of 1978 (Pub. L. 95–242)

- West Valley Demonstration Project Act of 1980 (Pub. L. 96–368)

- Nuclear Waste Policy Act of 1982, as amended (Pub. L. 97–425)

- Low-Level Radioactive Waste Policy Amendments Act of 1985 (Pub. L. 99–240)

- Energy Policy Act of 1992 (Pub. L. 102–486)

- Energy Policy Act of 2005 (Pub. L. 109-58)

The NRC, its licensees (those licensed by the NRC to use radioactive materials), and the Agreement States (States that assume regulatory authority over their own use of certain nuclear materials) share a common responsibility to protect public health and safety and the environment. Federal regulations and the NRC regulatory program are important elements in the protection of the public. However, because licensees are the ones using radioactive material, they bear the primary responsibility for safely handling and using these materials.

Major Activities

The NRC fulfills its responsibilities through the following licensing and regulatory activities:

- licenses the design, construction, operation, and decommissioning of commercial nuclear power plants and other nuclear facilities, such as uranium enrichment facilities, and research and test reactors;

- licenses the possession, use, processing, handling, and importing and exporting of nuclear materials;

- licenses the siting, design, construction, operation, and closure of low-level radioactive waste disposal sites in States under NRC jurisdiction;

- licenses the operators of nuclear reactors;

- inspects licensed and certified facilities and activities;

- certifies gaseous diffusion enrichment facilities and licences other enrichment facilities;

- conducts light-water reactor safety research, using independent research, data, and expertise, to develop regulations and anticipate potential safety problems;

- collects, analyzes, and disseminates information about the operational safety of commercial nuclear power reactors and certain nonreactor activities;

- issues safety and security regulations, policies, goals, and orders that govern licensed nuclear activities and interacts with other Federal agencies, including the U.S. Department of Homeland Security (DHS), on safety and security issues;

- investigates nuclear incidents and allegations concerning any matter regulated by the NRC;

- enforces NRC regulations and the conditions of NRC licenses and may impose civil sanctions, including civil penalties, for violations;

- conducts public hearings on matters of nuclear and radiological safety, environmental concern, and common defense and security;

- implements U.S. Government international legal commitments under treaties and conventions;

- develops effective working relationships with State and Tribal governments regarding reactor operations and the regulation of nuclear materials;

- directs the NRC program for response to incidents involving licensees and conducts a program of emergency preparedness and response for licensed nuclear facilities;

- evaluates and acts on the lessons learned from the March 11, 2011, nuclear accident in Japan to ensure that appropriate safety enhancements are implemented at U.S. commercial nuclear facilities;

- provides opportunities for public involvement in the regulatory process that include: holding open meetings, conferences, and workshops; soliciting public comments on proposed regulations, petitions, guidance documents, and draft technical reports; responding to requests for NRC documents under the Freedom of Information Act; reporting safety concerns; and providing access to hundreds of thousands of NRC documents through the NRC Web site; and

- participates in Open Government initiatives that focus on open, accountable, and accessible government and engage the public in dialogue and interactions, such as the use of social media and interactive high-value data sets.

The NRC Headquarters complex, located in Rockville, MD.

FY 2011 Accomplishment Highlights

Reactors

- completed all required inspection and assessment activities of the Reactor Oversight Process, including the initiation of 21 reactive inspections

- renewed 12 reactor licenses

- logged 1,300 reactor licensing tasks and activities

- collaborated with States, Federal agencies, and licensees in responding to significant natural events, including tornados, floods, hurricanes, and earthquakes

- initiated preconstruction inspections at Vogtle Units 3 and 4, and Summer Units 2 and 3

- monitored ongoing construction inspections at Watts Bar

- conducted safety and environmental reviews of the first two new reactor combined license applications and a mandatory hearing for the Vogtle application

- approved cyber security plans for all nuclear power plants

- completed significant emergency preparedness rulemaking

- processed National Fire Protection Association (NFPA) Standard 805 pilot applications and coordinated the review schedule for 29 upcoming NFPA 805 submittals

- responded to the Japan nuclear accident and began implementation of the lessons-learned initiative, including the Near-Term Task Force report

- hosted the third Integrated Regulatory Review Service lessons-learned workshop

- wrote the national report on the safety of U.S. nuclear power plants, "United States Fifth National Report for the Convention on Nuclear Safety"

- conducted the final phase 1 ignition (zirconium fire) test of the Sandia Fuel Project's single full-length insulated commercial pressurized-water reactor fuel assembly

- issued the Safety Culture Policy Statement (applicable to all licensees)

- participated in 18 new international agreements on cooperative research with other countries, adding to the existing 90 active agreements

- published extensive research results on a wide variety of topics to confirm the safety of operating facilities

Materials and Waste

- reviewed and approved three new uranium recovery licenses and the restart of one uranium recovery facility

- oversaw the safe construction of the URENCO LES enrichment facility and the Mixed Oxide Fuel Fabrication Facility

- implemented the License Tracking System, Version 2, and the National Source Tracking System, Version 2.2

- issued the final policy statement on the protection of sealed radiation sources containing cesium-137 chloride

- oversaw the orderly closure of the NRC's Yucca Mountain high-level waste repository licensing program

- conducted safety and environmental reviews and the mandatory hearing for licensing the AREVA Eagle Rock centrifuge enrichment facility in Idaho

- enhanced coordination with the States through the Integrated Materials Performance Evaluation Program and the State Liaison Officers

- made substantial progress on numerous rulemakings (including 10 CFR Parts 20, 35, 37, 40, 61, 73, and 74)

Corporate

- ensured that construction of Three White Flint North remained on schedule

- oversaw the One White Flint North lobby expansion and ongoing interior renovation

- implemented the new Financial Accounting and Integrated Management Information System

- improved the Human Resources Management System

- awarded the information technology support contract (the largest contract that the NRC has ever awarded)

- launched Agencywide Documents Access and Management System–Version P8

- held the 23rd Annual Regulatory Information Conference

- launched the new public Web site with improved navigation, content, and accessibility

- launched the use of social media at the NRC (external blog, Twitter, and YouTube)

- held 1,040 public meetings

- processed 381 Freedom of Information Act requests (closed 338)

- awarded $12.4 million in grants to 91 minority-serving (higher education) institutions

- received highest large-agency ranking in all four of Office of Personnel Management's key indices: leadership and knowledge management, performance culture, talent management, and job satisfaction

Organizations and Functions

The NRC's Commission consists of five members nominated by the President and confirmed by the U.S. Senate for 5-year terms. The President designates one member to serve as Chairman, principal executive officer, and spokesperson of the Commission. The members' terms are staggered so that one Commissioner's term expires on June 30 every year. No more than three Commissioners can belong to the same political party. The members of the Commission are listed below. The Commission as a whole formulates policies and regulations governing nuclear reactor and materials safety, issues orders to licensees, and adjudicates legal matters brought before it. The Executive Director for Operations carries out the policies and decisions of the Commission and directs the activities of the program and regional offices (see Figures 2 and 3).

Commissioner Term Expiration

Commissioner	Expiration of Term
Allison M. Macfarlane, Chairman	June 30, 2013
Kristine L. Svinicki	June 30, 2017
George Apostolakis	June 30, 2014
William D. Magwood, IV	June 30, 2015
William C. Ostendorff	June 30, 2016

The NRC has its headquarters in Rockville, MD, and maintains four regional offices in King of Prussia, PA; Atlanta, GA; Lisle, IL; and Arlington, TX. The NRC includes the major program offices described below:

Office of Nuclear Reactor Regulation handles all licensing and inspection activities associated with the operation of existing nuclear power reactors and research and test reactors.

Office of New Reactors provides safety oversight of the design, siting, licensing, and construction of new commercial nuclear power reactors.

Office of Nuclear Material Safety and Safeguards regulates activities that provide for the safe and secure production of nuclear fuel used in commercial nuclear reactors; the safe storage, transportation, and disposal of high- and low-level radioactive waste and spent nuclear fuel; and the transportation of radioactive materials regulated under the Atomic Energy Act of 1954, as amended.

Figure 2. NRC Organizational Chart

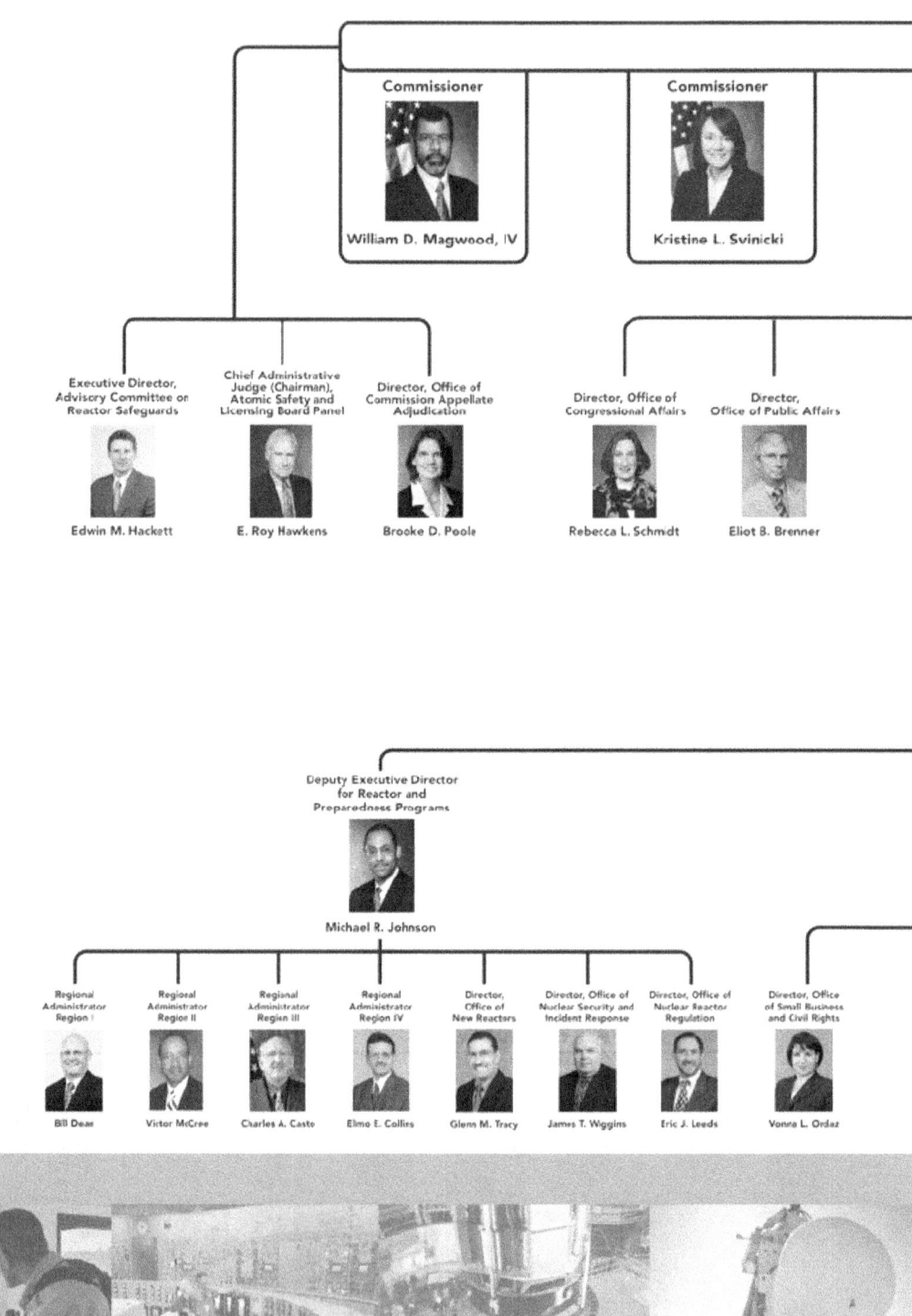

NRC: An Independent Regulatory Agency

Commissioner
William D. Magwood, IV

Commissioner
Kristine L. Svinicki

Executive Director, Advisory Committee on Reactor Safeguards
Edwin M. Hackett

Chief Administrative Judge (Chairman), Atomic Safety and Licensing Board Panel
E. Roy Hawkens

Director, Office of Commission Appellate Adjudication
Brooke D. Poole

Director, Office of Congressional Affairs
Rebecca L. Schmidt

Director, Office of Public Affairs
Eliot B. Brenner

Deputy Executive Director for Reactor and Preparedness Programs
Michael R. Johnson

Regional Administrator Region I
Bill Dean

Regional Administrator Region II
Victor McCree

Regional Administrator Region III
Charles A. Casto

Regional Administrator Region IV
Elmo E. Collins

Director, Office of New Reactors
Glenn M. Tracy

Director, Office of Nuclear Security and Incident Response
James T. Wiggins

Director, Office of Nuclear Reactor Regulation
Eric J. Leeds

Director, Office of Small Business and Civil Rights
Vonna L. Ordaz

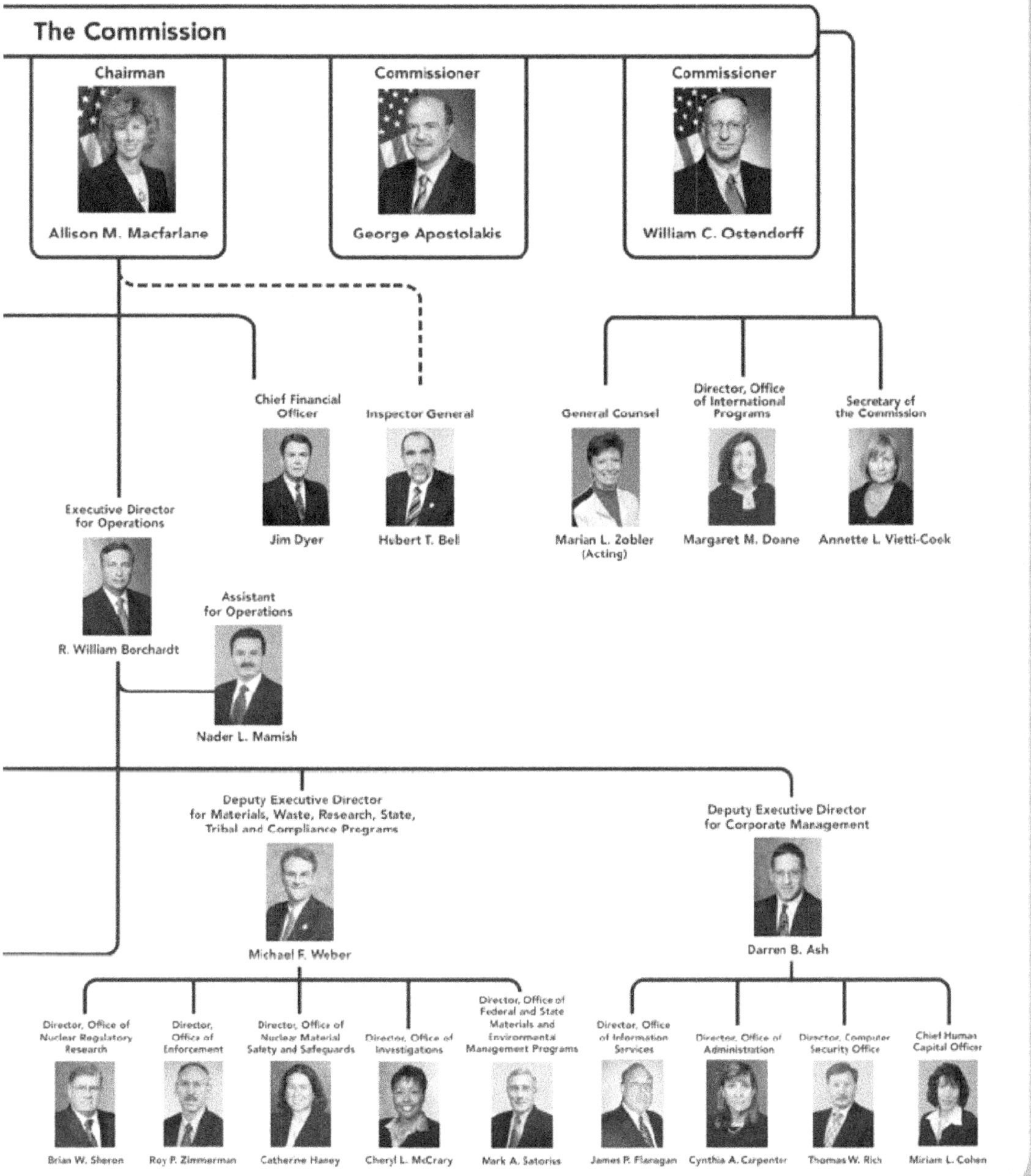

The Commission

Chairman

Allison M. Macfarlane

Commissioner

George Apostolakis

Commissioner

William C. Ostendorff

Chief Financial Officer

Jim Dyer

Inspector General

Hubert T. Bell

General Counsel

Marian L. Zobler
(Acting)

Director, Office of International Programs

Margaret M. Doane

Secretary of the Commission

Annette L. Vietti-Cook

Executive Director for Operations

R. William Borchardt

Assistant for Operations

Nader L. Mamish

Deputy Executive Director for Materials, Waste, Research, State, Tribal and Compliance Programs

Michael F. Weber

Deputy Executive Director for Corporate Management

Darren B. Ash

Director, Office of Nuclear Regulatory Research

Brian W. Sheron

Director, Office of Enforcement

Roy P. Zimmerman

Director, Office of Nuclear Material Safety and Safeguards

Catherine Haney

Director, Office of Investigations

Cheryl L. McCrary

Director, Office of Federal and State Materials and Environmental Management Programs

Mark A. Satorius

Director, Office of Information Services

James P. Flanagan

Director, Office of Administration

Cynthia A. Carpenter

Director, Computer Security Office

Thomas W. Rich

Chief Human Capital Officer

Miriam L. Cohen

Figure 3. NRC Regions

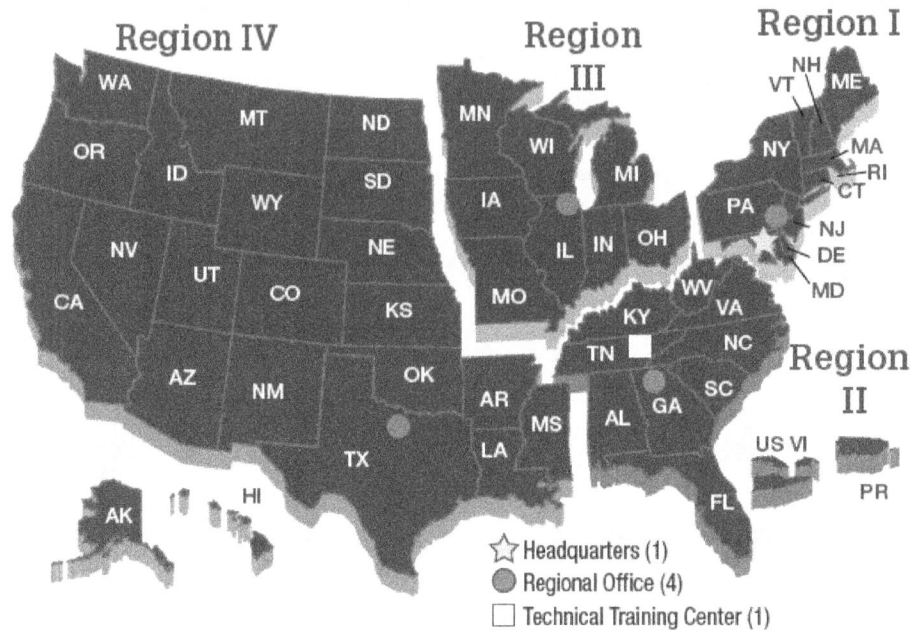

Region IV

Region III

Region I

Region II

☆ Headquarters (1)
● Regional Office (4)
☐ Technical Training Center (1)

Nuclear Power Plants
- Each regional office oversees the plants in its region except the Grand Gulf plant in Mississippi and the Callaway plant in Missouri, which Region IV oversees.

Materials Licensees
- Region I oversees licensees and Federal facilities located geographically in Region I and Region II.
- Region III oversees licensees and Federal facilities located geographically in Region III.
- Region IV oversees licensees and Federal facilities located geographically in Region IV.

Nuclear Fuel Processing Facilities
- Region II oversees all the fuel processing facilities in the region and those in Illinois, New Mexico, and Washington.
- Region II also handles all construction inspectors' activities for new nuclear power plants and fuel cycle facilities in all regions.

Office of Federal and State Materials and Environmental Management Programs develops and oversees the regulatory framework for the safe and secure use of nuclear materials; medical, industrial, academic, and commercial applications; uranium recovery activities; low-level radioactive waste sites; and the decommissioning of previously operating nuclear facilities and power plants. It works with Federal agencies, States, and Tribal and local governments on regulatory matters.

Office of Nuclear Regulatory Research provides independent expertise and information for making timely regulatory judgments, anticipating problems of potential safety significance, and resolving safety issues. It helps develop technical regulations and standards and collects, analyzes, and disseminates information about the operational safety of commercial nuclear power plants and certain nuclear materials activities.

Office of Nuclear Security and Incident Response oversees agency security policy for nuclear facilities and users of radioactive material. It provides a safeguards and security interface with other Federal agencies and maintains the agency's emergency preparedness and incident response program.

Regional Offices conduct inspection, enforcement (in conjunction with the Office of Enforcement), investigation, licensing, and emergency response programs for nuclear reactors, fuel facilities, and materials licensees.

Budget

For fiscal year (FY) 2012 (October 1, 2011–September 30, 2012), Congress appropriated $1.0381 billion ($1,038.1 million) to the NRC. The NRC's FY 2012 personnel ceiling is 3,953 full-time equivalent (FTE) staff (see Figures 4 and 5). The Office of the Inspector General received its own appropriation of $10.9 million. The amount is included in the total NRC budget. The breakdown of the budget is shown in Figure 5.

By law, the NRC must recover, through fees billed to licensees, approximately 90 percent of its budget authority for FY 2012, less the amounts appropriated from general funds for waste-incidental-to-reprocessing and generic homeland security activities. The NRC collects fees each year by September 30 and transfers them to the U.S. Treasury (see Figure 5). The total budget amount to be recovered by the NRC in FY 2012 is approximately $909.5 million.

Figure 4. NRC Budget Authority and Personnel Ceiling, FYs 2002–2012

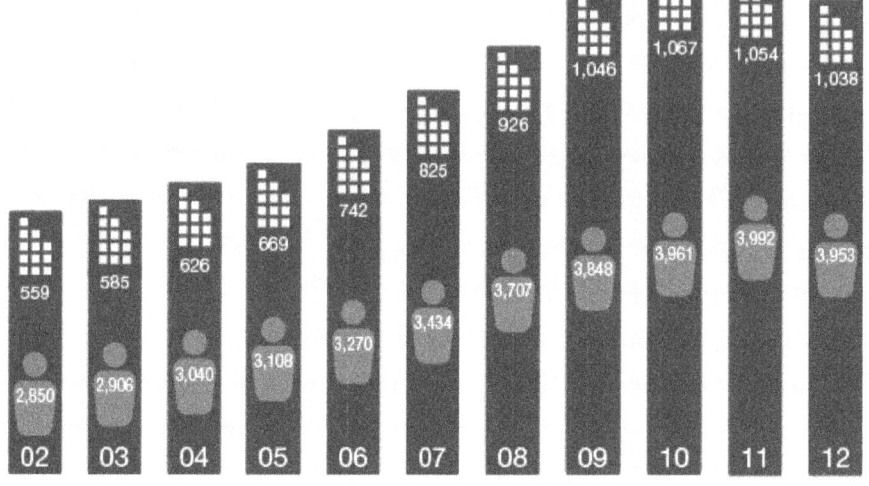

 Budget Authority
Dollars in Millions

Full-Time Equivalent (FTE) Staff

Note: Dollars are rounded to the nearest million.

Headquarters

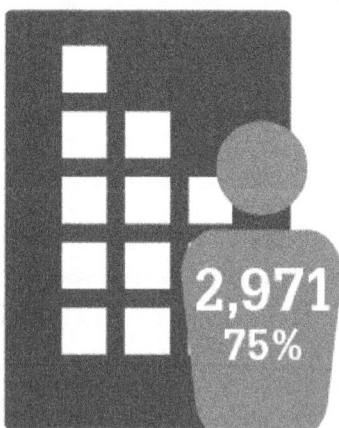

2,971
75%

FY 2012
Staff by Location
Total FTE: 3,953

Regions

982
25%

Figure 5. NRC FY 2012 Distribution of Budget Authority a d Staff
Recovery of NRC Budget

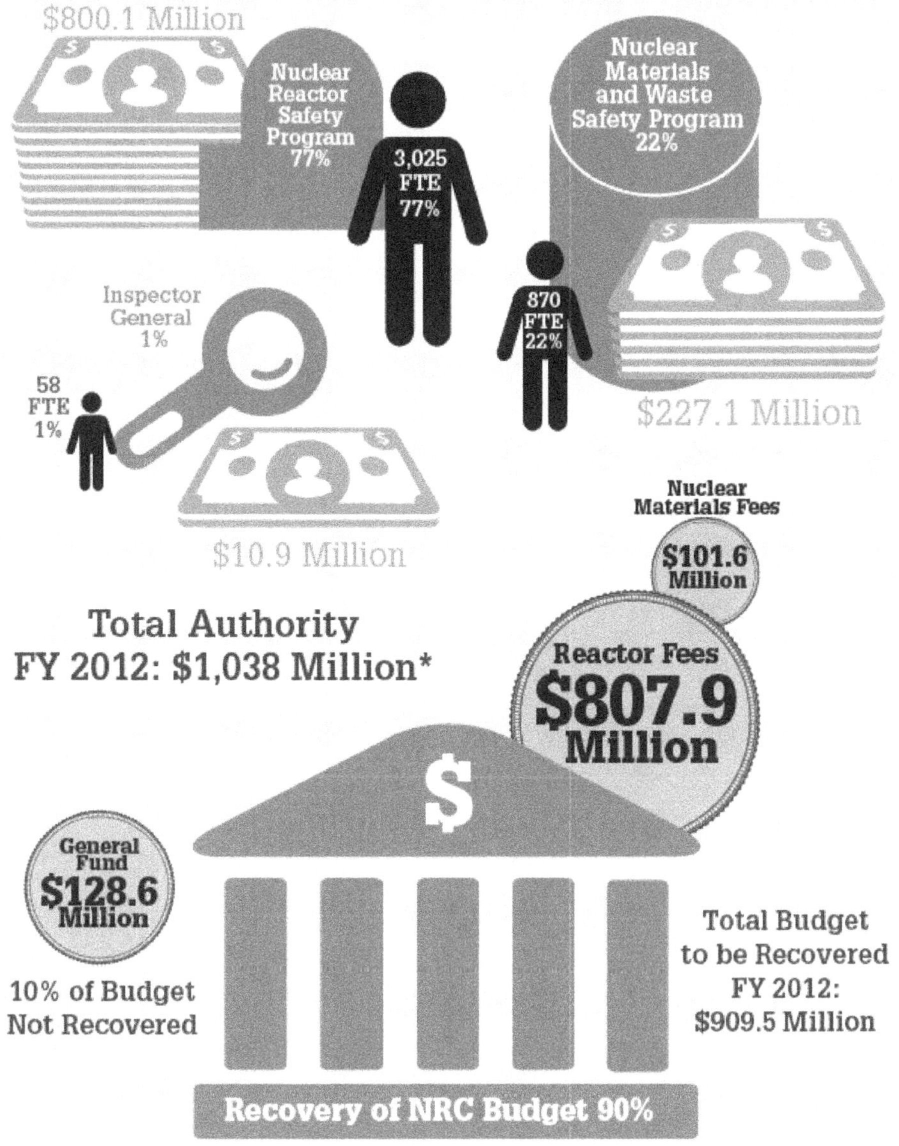

$800.1 Million

Nuclear
Reactor
Safety
Program
77%

3,025
FTE
77%

Nuclear
Materials
and Waste
Safety Program
22%

870
FTE
22%

$227.1 Million

Inspector
General
1%

58
FTE
1%

$10.9 Million

Total Authority
FY 2012: $1,038 Million*

Nuclear
Materials Fees

$101.6
Million

Reactor Fees

$807.9
Million

General
Fund
$128.6
Million

10% of Budget
Not Recovered

Total Budget
to be Recovered
FY 2012:
$909.5 Million

Recovery of NRC Budget 90%

Note: The NRC incorporates corporate and administrative costs proportionately within programs.

U.S. and Worldwide Nuclear Energy

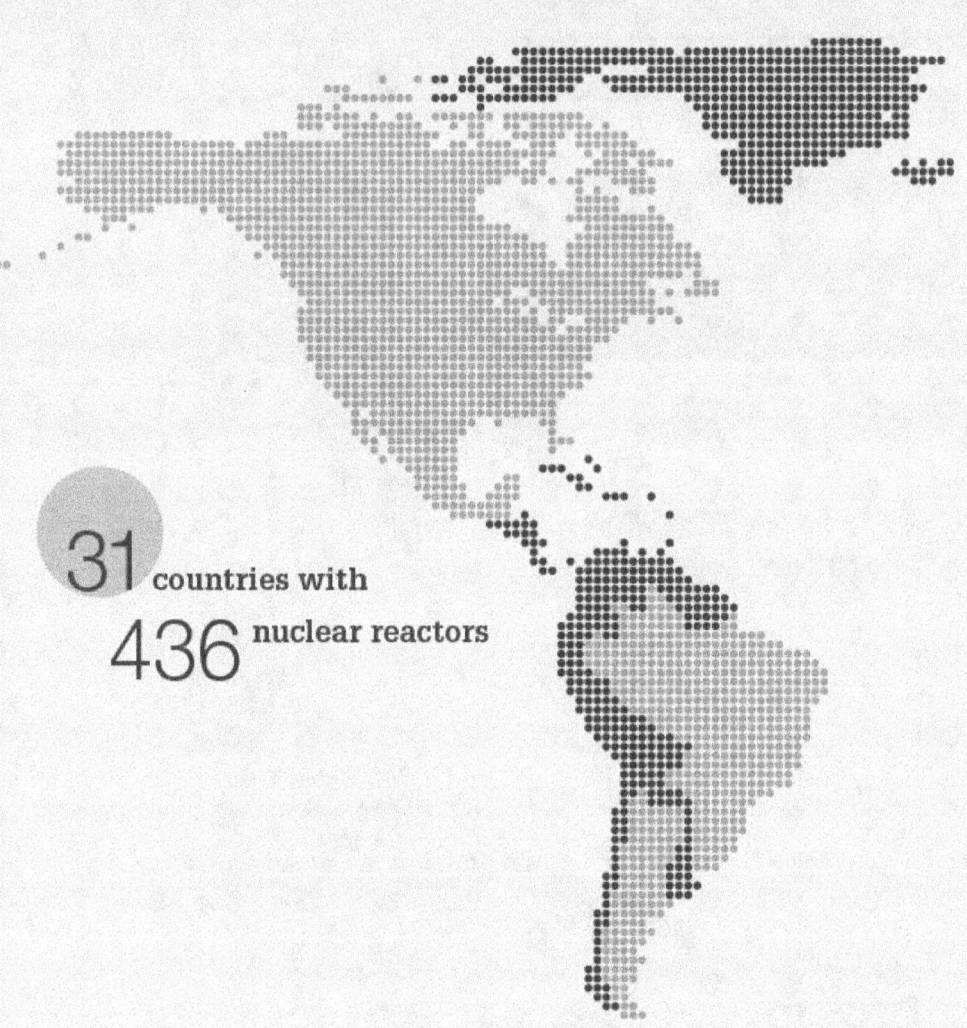

31 countries with

436 nuclear reactors

Nuclear Share Generated

France
78%

Belgium
54%

Slovakia
54%

Ukraine
47%

Hungary
43%

Slovenia
42%

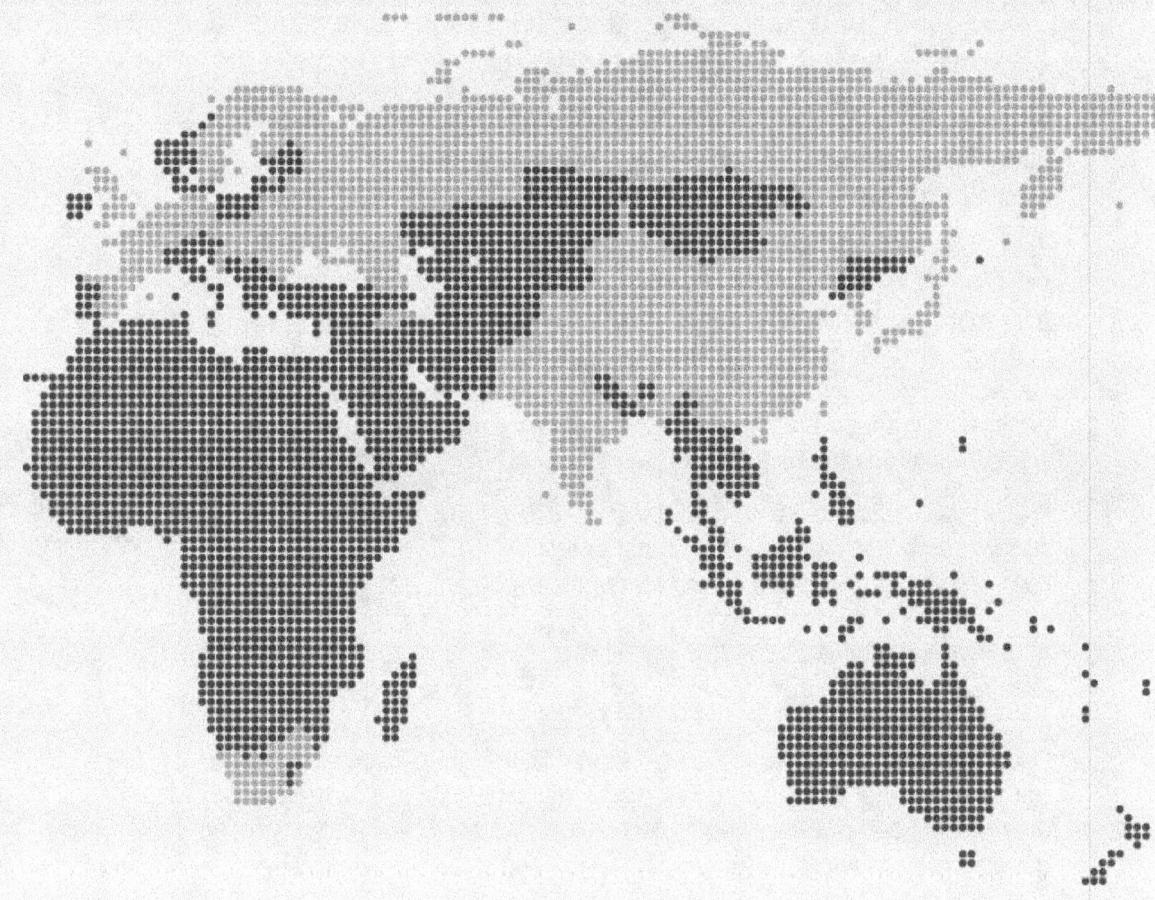

Switzerland	Sweden	Rep Korea	Bulgaria	Spain	U.S.A.

41%	40%	35%	33%	20%	19%

U.S. Electricity Generated by Commercial Nuclear Power

As of May 2012, the 104 NRC-licensed nuclear reactors accounted for about 20 percent of U.S. net electric generation, providing 790 billion kilowatthours of electricity (see Figures 6 and 7).

Thirty-one of the 50 States generate electricity from nuclear power plants. Of these states, three (New Jersey, South Carolina, and Vermont) generated more than 50 percent of their electricity from nuclear power. In addition, 12 States generated 25 to 50 percent of their electricity from nuclear power. The data cited reflect the percentages of the total net generation in these States that were from nuclear sources (see Figure 8).

See Appendix L for the nuclear electricity generated by State.

Since the 1970s, the Nation's utilities have sought power uprates as a way to generate more electricity from existing nuclear plants. By January 2012, the NRC had approved 140 power uprates, resulting in a gain of approximately 6,194 megawatts electric (MWe) at existing plants. Collectively, these uprates have added the equivalent of six new reactors' worth of electrical generation at existing plants. Licensees responding to a December 2011 NRC survey indicated that they plan to submit 15 power uprate applications in the next 5 years. If these applications are approved, the resulting uprates would add another 1,160 MWe to the Nation's generating capacity (see Figure 9).

Worldwide Electricity Generated by Commercial Nuclear Power

As of May 2012, there were 436 operating reactors (at least partially) in 31 countries with a total installed capacity of 370,499 megawatts electric (MWe) (see Figure 10). In addition, five nuclear power plants were in long-term shutdown, and 66 were under construction. Based on preliminary data in 2011, France had the highest nuclear portion (78 percent) of total domestic energy generated (see Figure 11).

Figure 6. U.S. Net Electric Generation by Energy Source, 2011

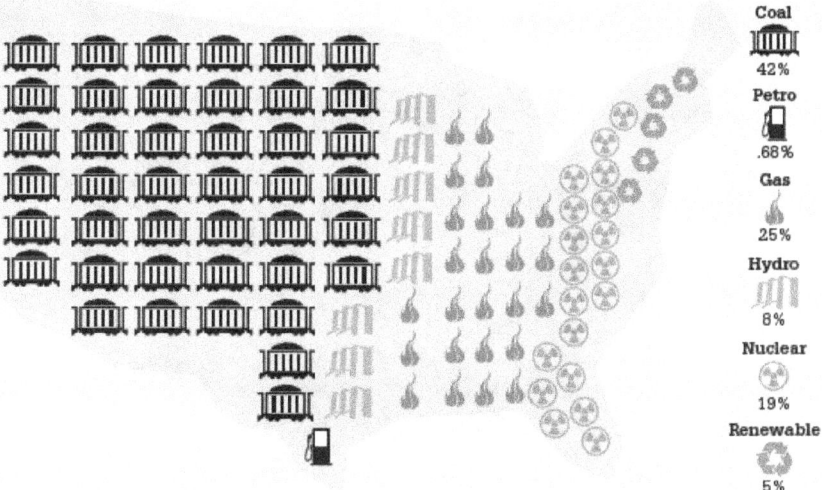

Coal
42%

Petro
.68%

Gas
25%

Hydro
8%

Nuclear
19%

Renewable
5%

Source: DOE/EIA, May 2012, www.eia.doe.gov

Figure 7. U S. Net Electric Generation by Ene gy Source, 2002–2011

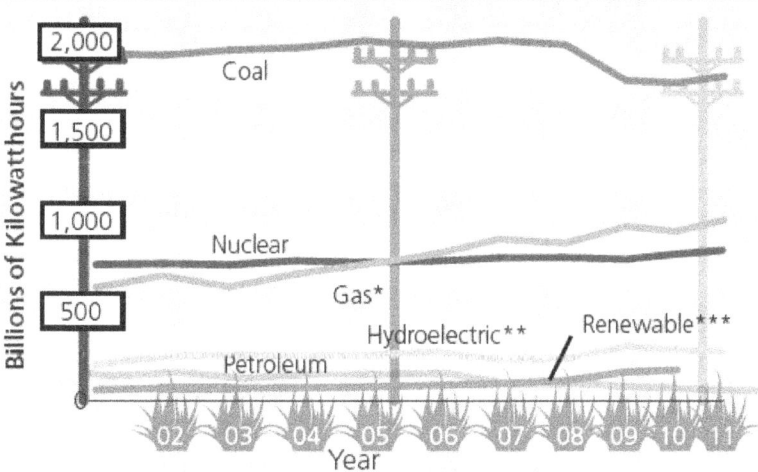

* Gas includes natural gas, blast furnace gas, propane gas, and other manufactured and waste gases derived from fossil fuel.
** Hydroelectric includes conventional hydroelectric and hydroelectric pumped storage.
*** Renewable energy includes geothermal, wood and nonwood waste, wind, and solar energy.
Source: DOE/EIA, May 2012, www.eia.doe.gov

Figure 8. Net Electricity Generated in Each State by Nuclear Power

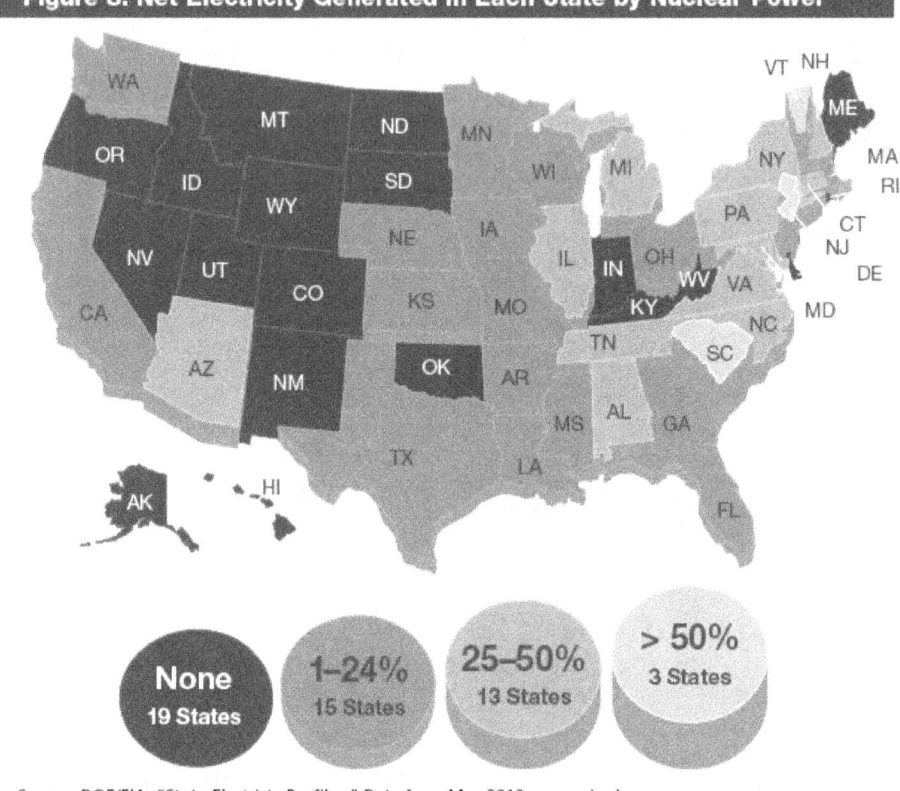

Source: DOE/EIA, "State Electricty Profiles," Data from May 2012, www.eia.doe.gov

Figure 9. Power Uprates: Past, Current, and Future

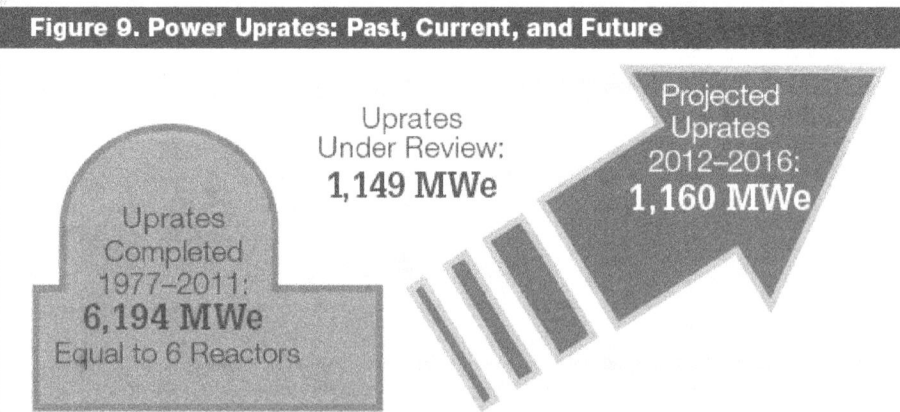

Uprates Under Review: 1,149 MWe

Projected Uprates 2012–2016: 1,160 MWe

Uprates Completed 1977–2011: 6,194 MWe Equal to 6 Reactors

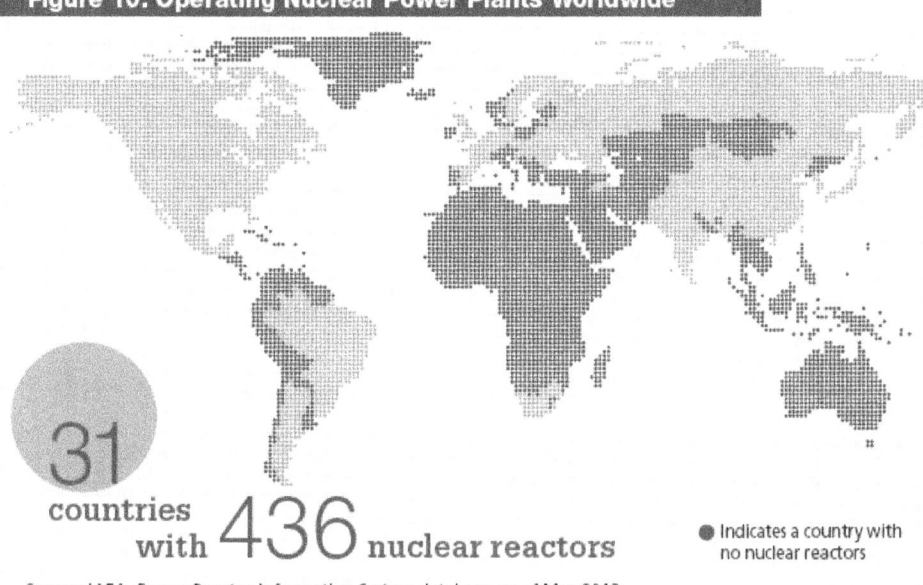

Figure 10. Operating Nuclear Power Plants Worldwide

31 countries with 436 nuclear reactors

● Indicates a country with no nuclear reactors

Source: IAEA, Power Reactor Information System database, as of May 2012

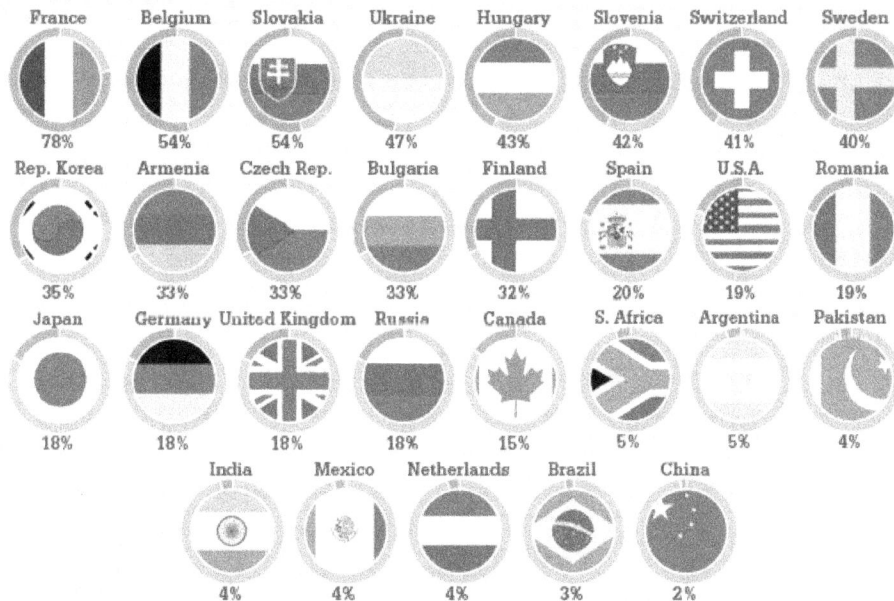

Figure 11. Nuclear Share of Electricity Generated by Country, 2011

France	Belgium	Slovakia	Ukraine	Hungary	Slovenia	Switzerland	Sweden
78%	54%	54%	47%	43%	42%	41%	40%

Rep. Korea	Armenia	Czech Rep.	Bulgaria	Finland	Spain	U.S.A.	Romania
35%	33%	33%	33%	32%	20%	19%	19%

Japan	Germany	United Kingdom	Russia	Canada	S. Africa	Argentina	Pakistan
18%	18%	18%	18%	15%	5%	5%	4%

India	Mexico	Netherlands	Brazil	China
4%	4%	4%	3%	2%

Note: The country's short-form name is used.
Source: IAEA, Power Reactor Information System database, as of May 2012

International Activities

The NRC performs certain legislatively mandated international duties. These include licensing the import and export of nuclear materials and equipment and participating in activities supporting U.S. Government compliance with international treaty and agreement obligations. The NRC has bilateral programs of assistance or cooperation with 42 countries, Taiwan, and the European Atomic Energy Community (see Figure 12).

See Appendix S for the number of nuclear power reactors by nation and Appendix T for nuclear power units by reactor type, worldwide.

The NRC has supported U.S. Government nuclear safety initiatives with countries in Europe, Africa, Asia, and Latin America. In addition, the NRC actively cooperates with multinational organizations, such as the International Atomic Energy Agency (IAEA) and the Nuclear Energy Agency (NEA), a part of the Organization for Economic Co-operation and Development. The NRC also has a robust international cooperative research program.

Since its inception, the agency has hosted over 350 foreign nationals in on-the-job training assignments at NRC Headquarters and the regional offices. The NRC's Foreign Assignee Program helps instill regulatory awareness, capabilities, and commitments in exchanges with assignees from other countries. It also helps to enhance the regulatory expertise of both foreign assignees and NRC staff. Additionally, the program improves channels of communication through interaction with the international nuclear community and development of relationships with key personnel in foreign regulatory agencies. Through its export and import authority, the NRC upholds the U.S. Government goals of limiting the proliferation of materials that could be used in weapons, and supports the safe and secure use of civilian nuclear and radioactive materials worldwide. The NRC continues to work to strengthen the export and import regulations of nuclear equipment and materials and to improve communication between domestic and international stakeholders.

The NRC assists in implementing the U.S. Government's international nuclear policies through developing and implementing legal instruments that address nuclear nonproliferation, safety, international safeguards, physical protection, emergency notification and assistance, spent fuel and waste management, and liability.

Figure 12. Bilateral Information Exchange and Cooperation Agreements with the NRC

Agreement Country, Renewal Date

Argentina, 2012	Germany, 2012	Poland, 2015
Armenia, 2012	Greece, 2013	Romania, 2016
Australia, 2013	Hungary, 2012	Russia*, 2001
Belgium, 2014	Indonesia, 2013	Slovakia, 2015
Brazil, 2014	Israel, 2016	Slovenia, 2015
Bulgaria*, 2011	Italy, 2015	South Africa, 2014
Canada, 2012	Japan, 2015	Spain, 2015
China, 2013	Kazakhstan, 2014	Sweden*, 2011
Croatia, 2013	Korea, Rep. of, 2015	Switzerland, 2012
Czech Republic, 2014	Lithuania, 2015	Thailand, 2012
Egypt, 1991	Mexico, 2012	Ukraine, 2016
EURATOM, 2014	Netherlands, 2013	United Arab Emirates, 2015
Finland*, 2011	Peru, Open-Ended	United Kingdom, 2013
France, 2013	Philippines, Open-Ended	Vietnam, 2013

Note: The country's short-form name is used. The NRC also provides support to the American Institute in Taiwan. Egypt's agreement has been deferred until its regulatory body requests reinstatement.

EURATOM—The European Atomic Energy Community

* In negotiation

The NRC participates in the negotiation and implementation of U.S. bilateral agreements for peaceful nuclear cooperation under Section 123 of the U.S. Atomic Energy Act of 1954, as amended. The NRC ensures licensee compliance with the U.S. Voluntary Safeguards Offer agreement and its additional protocol to the U.S.-IAEA Agreement for the Application of Safeguards in the United States.

The NRC also participates in a wide range of mutually beneficial international exchange programs that enhance the safety and security of peaceful nuclear activities worldwide. These low-cost, high-impact programs provide joint cooperative activities and assistance to other countries to develop and improve regulatory organizations. The NRC engages in the following activities:

- cooperates with countries with mature nuclear programs to ensure the timely exchange of applicable nuclear safety and security information relating to operating reactors and consults with these countries on new reactor-related activities;

- ensures prompt notification to foreign partners about U.S. safety issues, notifies NRC program offices about foreign safety issues, and shares security information with selected countries;

- initiates bilateral discussions in such regulatory areas as licensing, inspection, and enforcement with countries that have recently built facilities or have vendors of equipment that may be imported to the United States during the anticipated construction of new nuclear power plants;

Photo courtesy of IAEA

The NRC participates in the annual General International Conference for the IAEA in Vienna, Austria.

- participates in the Multinational Design Evaluation Program, which leverages the resources of interested regulatory authorities to review new designs of nuclear power reactors;

- participates in a variety of conventions, treaties, and other legal and political instruments that together make up the international nuclear regime. For example, the NRC participated in the Third Review Meeting of the Joint Convention on the Safety of Spent Fuel Management and on the Safety of Radioactive Waste Management in May 2012, and the Extraordinary Meeting of the Convention on Nuclear Safety, convened specifically to address followup to Fukushima-related issues, in August 2012. The NRC also provided technical and policy support to the U.S. delegation to the 2012 Preparatory Committee Meeting of the Nuclear Non-Proliferation Treaty;

- provides guidance about export and import licensing for nuclear materials and equipment published in 10 CFR Part 110, "Export and Import of Nuclear Equipment and Material"; the NRC continues its outreach to other countries on the Code of Conduct on the Safety and Security of Radioactive Sources and through bilateral meetings to ensure consistency in national regulatory approaches;

- assists other countries in developing and improving regulatory programs through training, workshops, peer review of regulatory documents, working group meetings, and exchanges of technical information and specialists;

- assists countries, through a pilot program begun in 2008, to ensure regulatory control over radioactive sources through development of standards and provision of training and workshops; in 2010, the program expanded to Latin America; and in 2011, outreach began to countries in Africa;

- participates in the programs of IAEA, NEA, and the European Union concerned with safety research and regulatory matters, radiation protection, risk assessment, emergency preparedness, waste management, transportation, safeguards, physical protection, security, standards development, training, technical assistance, and communications;

- participates in the International Nuclear Regulators Association meetings to influence and enhance nuclear safety. Association members are the most senior officials of well-established independent national nuclear regulatory organizations. Current members are Canada, France, Germany, Japan, Republic of Korea, Spain, Sweden, the United Kingdom, and the United States;

- meets, through the NRC's Advisory Committee on Reactor Safeguards (ACRS), with other international advisory committees through annual working group meetings and plenary meetings every 4 years to exchange information;

- participates in joint cooperative research programs through approximately 100 multilateral agreements with 30 countries and Taiwan to leverage access to foreign test facilities not otherwise available to the United States. Access to foreign test facilities expands the NRC's knowledge base and contributes to the efficient and effective use of the NRC's resources in conducting research on high-priority safety issues; and

- in October 2011, the NRC hosted an IAEA workshop on lessons learned from Integrated Regulatory Review Service (IRRS) missions, the purpose of which was for countries that have had IRRS missions to share experiences and insights to strengthen the IRRS process.

Immediately after the March 11, 2011, earthquake and tsunami in Japan, a team of subject matter experts on reactor safety, protective measures, and international relations from the NRC, the U.S. Department of Energy, and the U.S. Department of Health and Human Services traveled to Japan to help the Government of Japan assess and address the emergency at the Fukushima Dai-ichi nuclear power plant.

The NRC continues to maintain its longstanding relationship with its Japanese regulatory and other governmental and private sector counterparts. The agencies exchange technical information and lessons being learned as a result of the accident. Japanese counterparts include organizations such as the Nuclear and Industrial Safety Agency, the Japan Nuclear Energy Safety; Tokyo Electric Power Company; the Ministry of Economy, Trade and Industry; the Ministry of Education, Culture, Sports, Science and Technology; and the Ministry of Foreign Affairs.

In late 2011, the NRC ended its staff-level presence in Tokyo. The NRC Near-Term Task Force reviewed the events in Japan and has issued its findings and made recommendations for improvements to NRC requirements, programs, and processes. The Commission issued orders based on Task Force recommendations in March 2012. A steering committee will be established for the NRC and its Japanese regulatory counterpart to continue its information exchanges.

Figure 13. Actions in Response to the Japan Nuclear Accident: Timeline

March 11, 2011 (AM)

A magnitude 9.0 earthquake strikes near Honshu, Japan, generating an estimated 45-foot (14 meter) tsunami at the Fukushima Dai-ichi nuclear power plant.

Commission Public Meetings

The Commission briefs Congress and provides opportunities for citizens to be heard starting in March 2011.

April and May 2011

The NRC reports all U.S. nuclear power plants have appropriate post-9/11 emergency equipment and procedures in place.

September 2011

NRC resident inspectors begin examining U.S. nuclear fuel cycle facilities' plans and procedures for safely dealing with severe events.

March 11, 2011 (PM)

The NRC staffs its Headquarters Operations Center on a 24/7 basis, first monitoring tsunami effects on the U.S. West Coast, and then supporting the U.S. response to the Japan nuclear accident until May 16th, 2011. The first of many reactor experts are sent to Japan as part of a USAID mission.

March 23, 2011

NRC resident inspectors begin reexamining post-9/11 emergency equipment and related items at U.S. nuclear power plants, in light of details from the Fukushima accident.

July 2011

The NRC's Near-Term Task Force issues its report on lessons learned from Fukushima.

Next Steps

Over the next months, the NRC takes numerous actions on the lessons learned to ensure appropriate enhancements are implemented.

Nuclear Reactors

NRC Reactor Inspection Efforts

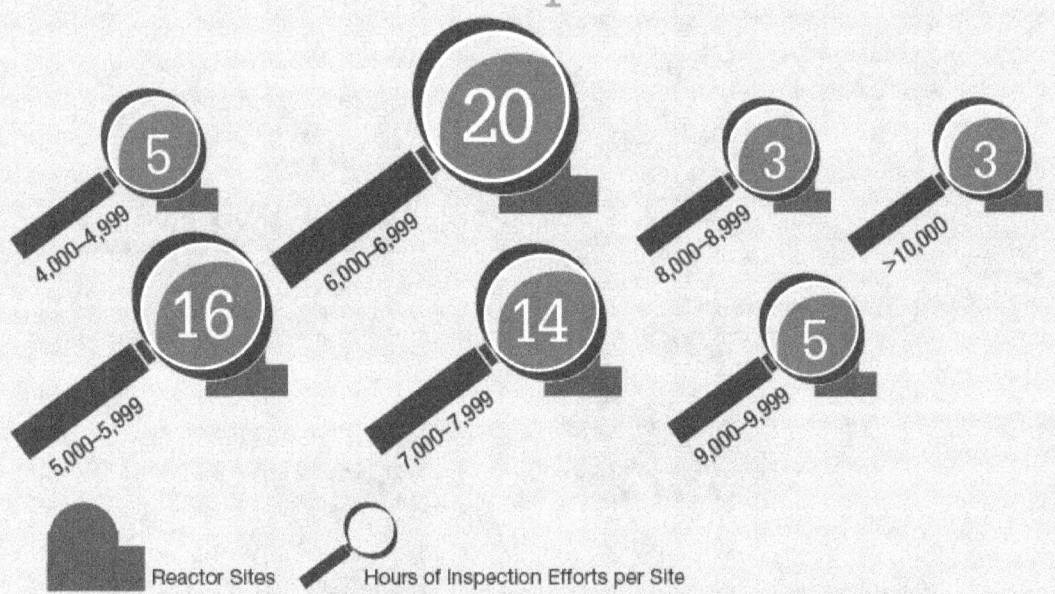

5
4,000–4,999

20
6,000–6,999

3
8,000–8,999

3
>10,000

16
5,000–5,999

14
7,000–7,999

5
9,000–9,999

Reactor Sites   Hours of Inspection Efforts per Site

1,500 MWt

SMALLEST COMMERCIAL POWER REACTOR

1,500 Megawatts thermal

20 MWt

LARGEST RESEARCH & TEST REACTOR

75x Smaller

20 Megawatts thermal

NRC Research Funding FY 2012

Total $49.8 Million

- Reactor Program—$42.8 M
- New/Advanced Reactor Licensing—$3.7 M
- Homeland Security—$1.5 M
- Materials and Waste—$1.3 M
- Infrastructure Support—$0.4 M

Expirations of Reactor Operating Licenses

2013–2018
4

2019–2022
5

2023–2030
27

2031–2049
68

U.S. Commercial Nuclear Power Reactors

As of August 2012, 104 commercial nuclear power reactors were licensed to operate in 31 States (see Figure 14). These reactors have the following characteristics:

- 4 different reactor vendors
- 26 operating companies
- 80 different designs
- 65 sites

See Appendix A for a listing of reactors and their general licensing information and Appendix U for Native American Reservations and Trust lands near nuclear power plants.

Diversity

Although there are many similarities, each reactor design can be considered unique. Figure 15 shows a typical pressurized-water reactor (PWR), and Figure 16 shows a typical boiling-water reactor (BWR). Currently there are 35 BWR and 69 PWR reactor designs.

Resident Inspectors

The NRC has at least two full-time inspectors at each nuclear power plant site to ensure that facilities are meeting NRC regulations.

Photo courtesy of Dominion Nuclear Connecticut, Inc.

Millstone Power Station, located in Waterford, CT.

Figure 14. U.S. Operating Commercial Nuclear Power Reactors

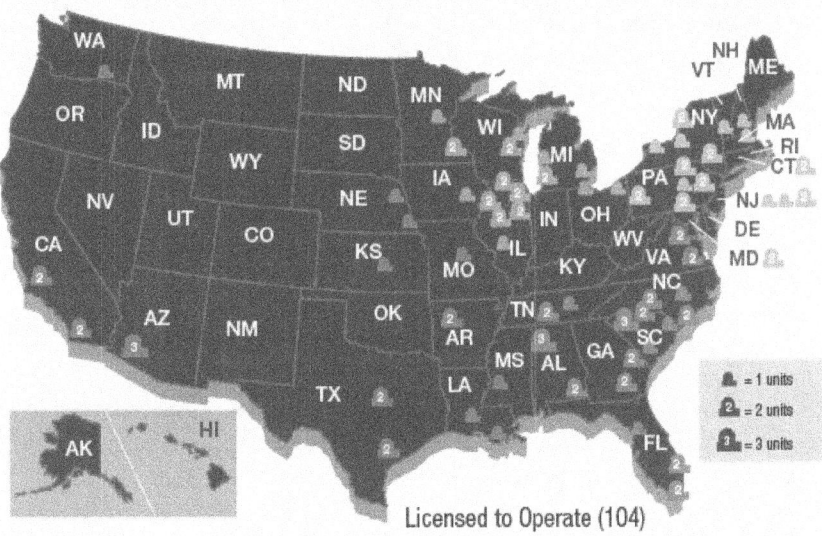

Licensed to Operate (104)

= 1 units
= 2 units
= 3 units

REGION I

CONNECTICUT
- Millstone 2 and 3

MARYLAND
- Calvert Cliffs 1 and 2

MASSACHUSETTS
- Pilgrim

NEW HAMPSHIRE
- Seabrook

NEW JERSEY
- Hope Creek
- Oyster Creek
- Salem 1 and 2

NEW YORK
- FitzPatrick
- Ginna
- Indian Point 2 and 3
- Nine Mile Point 1 and 2

PENNSYLVANIA
- Beaver Valley 1 and 2
- Limerick 1 and 2
- Peach Bottom 2 and 3
- Susquehanna 1 and 2
- Three Mile Island 1

VERMONT
- Vermont Yankee

REGION II

ALABAMA
- Browns Ferry 1, 2, and 3
- Farley 1 and 2

FLORIDA
- Crystal River 3
- St. Lucie 1 and 2
- Turkey Point 3 and 4

GEORGIA
- Edwin I. Hatch 1 and 2
- Vogtle 1 and 2

NORTH CAROLINA
- Brunswick 1 and 2
- McGuire 1 and 2
- Harris 1

SOUTH CAROLINA
- Catawba 1 and 2
- Oconee 1, 2, and 3
- Robinson 2
- Summer

TENNESSEE
- Sequoyah 1 and 2
- Watts Bar 1

VIRGINIA
- North Anna 1 and 2
- Surry 1 and 2

REGION III

ILLINOIS
- Braidwood 1 and 2
- Byron 1 and 2
- Clinton
- Dresden 2 and 3
- LaSalle 1 and 2
- Quad Cities 1 and 2

IOWA
- Duane Arnold

MICHIGAN
- Cook 1 and 2
- Fermi 2
- Palisades

MINNESOTA
- Monticello
- Prairie Island 1 and 2

OHIO
- Davis-Besse
- Perry

WISCONSIN
- Kewaunee
- Point Beach 1 and 2

REGION IV

ARKANSAS
- Arkansas Nuclear 1 and 2

ARIZONA
- Palo Verde 1, 2, and 3

CALIFORNIA
- Diablo Canyon 1 and 2
- San Onofre 2 and 3

KANSAS
- Wolf Creek 1

LOUISIANA
- River Bend 1
- Waterford 3

MISSISSIPPI
- Grand Gulf

MISSOURI
- Callaway

NEBRASKA
- Cooper
- Fort Calhoun

TEXAS
- Comanche Peak 1 and 2
- South Texas Project 1 and 2

WASHINGTON
- Columbia

Figure 15. Typical Pressurized-Water Reactor

How Nuclear Reactors Work

In a typical design concept of a commercial PWR, the following process occurs:

1. The core inside the reactor vessel creates heat.
2. Pressurized water in the primary coolant loop carries the heat to the steam generator.
3. Inside the steam generator, heat from the primary coolant loop vaporizes the water in a secondary loop, producing steam.
4. The steamline directs the steam to the main turbine, causing it to turn the turbine generator, which produces electricity.

The unused steam is exhausted to the condenser, where it is condensed into water. The resulting water is pumped out of the condenser with a series of pumps, reheated, and pumped back to the steam generator. The reactor's core contains fuel assemblies that are cooled by water circulated using electrically powered pumps. These pumps and other operating systems in the plant receive their power from the electrical grid. If offsite power is lost, emergency cooling water is supplied by other pumps, which can be powered by onsite diesel generators. Other safety systems, such as the containment cooling system, also need electric power. PWRs contain between 150–200 fuel assemblies.

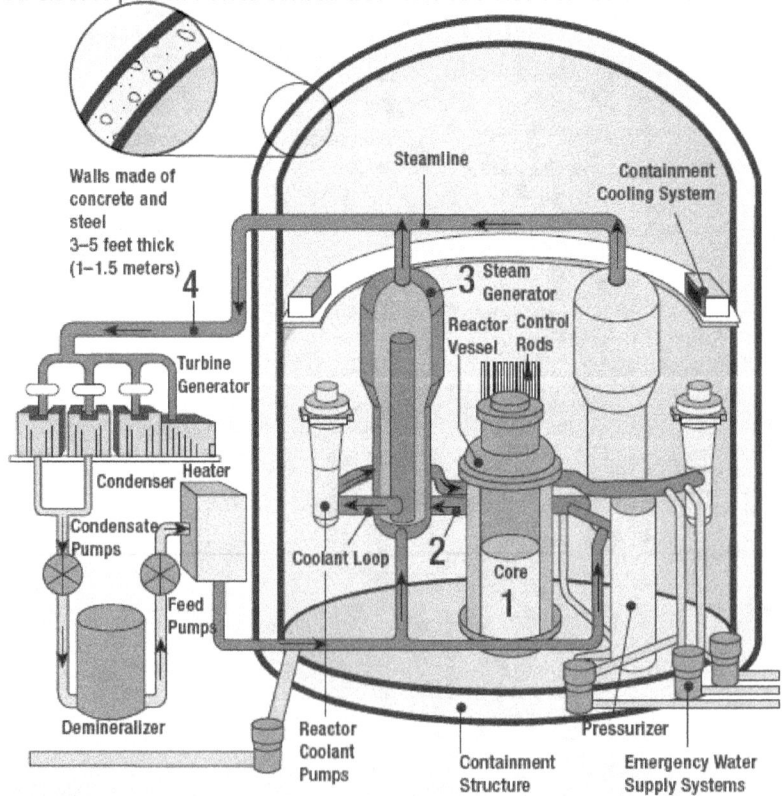

Figure 16. Typical Boiling-Water Reactor

How Nuclear Reactors Work

In a typical design concept of a commercial BWR, the following process occurs:

1. The core inside the reactor vessel creates heat.
2. A steam-water mixture is produced when very pure water (reactor coolant) moves upward through the core, absorbing heat.
3. The steam-water mixture leaves the top of the core and enters the two stages of moisture separation where water droplets are removed before the steam is allowed to enter the steamline.
4. The steamline directs the steam to the main turbine, causing it to turn the turbine generator, which produces electricity.

The unused steam is exhausted to the condenser, where it is condensed into water. The resulting water is pumped out of the condenser with a series of pumps, reheated, and pumped back to the reactor vessel. The reactor's core contains fuel assemblies that are cooled by water circulated using electrically powered pumps. These pumps and other operating systems in the plant receive their power from the electrical grid. If offsite power is lost, emergency cooling water is supplied by other pumps, which can be powered by onsite diesel generators. Other safety systems, such as the containment cooling system, also need electric power. BWRs contain between 370–800 fuel assemblies.

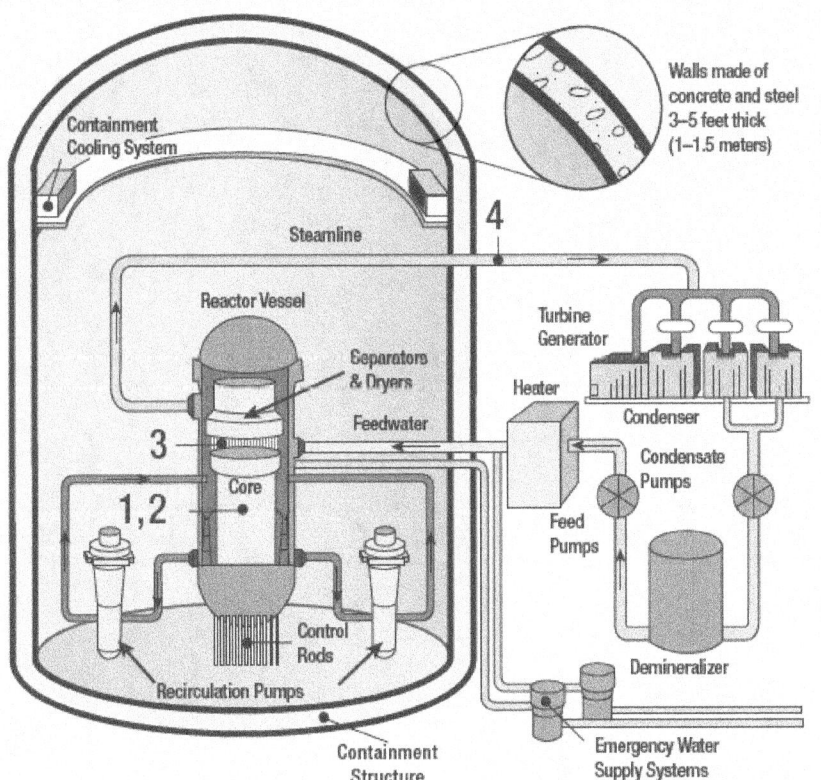

Figure 17. NRC Inspection Effort at Operating Reactors, 2011

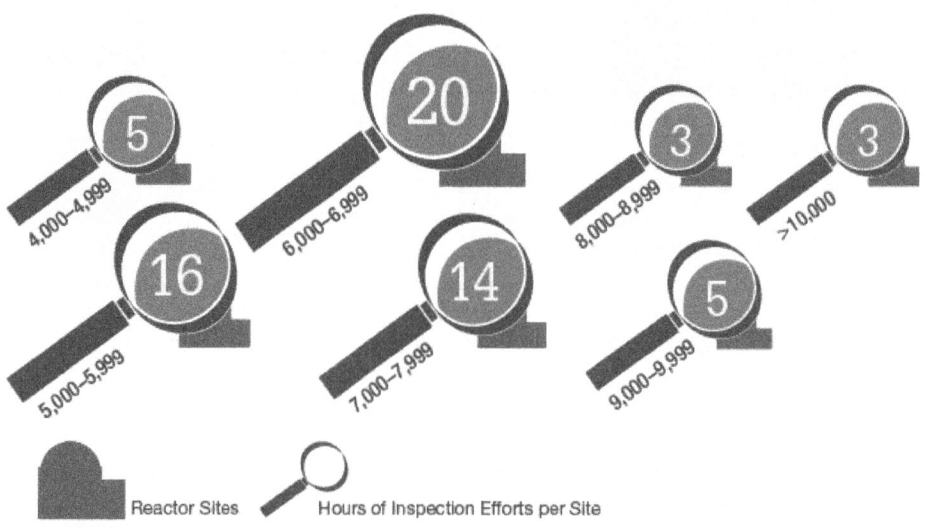

5
4,000–4,999

20
6,000–6,999

3
8,000–8,999

3
>10,000

16
5,000–5,999

14
7,000–7,999

5
9,000–9,999

Reactor Sites   Hours of Inspection Efforts per Site

Note: Data include Calendar Year (CY) 2011 hours for all activities related to baseline, plant-specific, generic safety issue, and allegation inspections.

* 66 total sites (including Indian Point Units 2 and 3, which are treated as separate sites for inspection effort)

An NRC inspector conducts routine inspections of plant equipment to ensure the plant is meeting NRC regulations.

Principal Licensing, Inspection, and Enforcement Activities

The NRC conducts a variety of licensing and inspection activities:

- The NRC is reviewing an operating license application from the Tennessee Valley Authority for the Watts Bar Unit 2 reactor under construction near Spring City, TN.

> *See Appendix B for permanently shut down and decommissioning reactors and Appendix W for significant enforcement actions.*

- Typically, each power reactor licensee requests about 10 separate license changes each year. The NRC completed more than 1,000 separate reviews in FY 2011.

- Currently, there are approximately 4,600 NRC-licensed reactor operators. Each operator must requalify every 2 years and apply for license renewal every 6 years.

- On average, the NRC expended approximately 6,820 hours of inspection-related effort at each operating reactor site during 2011 (see Figure 17).

- The NRC reviews applications for proposed new reactors and is developing an inspection program to oversee construction.

- The NRC reviews approximately 3,000 operating experience items, such as fire protection and access authorization programs, from licensed facilities annually.

- The NRC issues about 15 to 20 escalated enforcement actions per year to operating reactors for violations having a relatively high level of significance with regard to licensed activities affecting public health and safety. The primary enforcement actions, depending on the severity, are notices of violation, civil penalties, and orders.

- The NRC reviews approximately 600 allegations per year; allegations are assertions of inadequacy or impropriety associated with NRC-regulated activities.

- ACRS, an independent body of nuclear, engineering, and safety experts appointed by the Commission, reviews numerous safety issues for existing or proposed reactors and provides independent technical advice to the Commission. ACRS held 11 full Committee meetings and approximately 70 subcommittee meetings during 2011.

- The NRC currently oversees the decommissioning of 14 nuclear power reactors.

Oversight of U.S. Commercial Nuclear Power Reactors

The NRC does not operate nuclear power plants. Rather, it regulates the operation of the Nation's 104 nuclear power plants by establishing regulatory requirements for their design, construction, and operation. To ensure that the plants are operated safely within these requirements, the NRC licenses the plants to operate, licenses the plant operators, establishes technical specifications for the operation of each plant, and inspects plants daily.

Reactor Oversight Process

The NRC provides continuous oversight of plants through its Reactor Oversight Process (ROP) to verify that they are being operated in accordance with NRC rules, regulations, and license requirements. The NRC has full authority to take action to protect public health and safety, up to and including shutting a plant down.

In general terms, the ROP uses both NRC inspection findings and performance indicators from licensees to assess the safety performance and security measures of each plant. There are five levels that range from "fully meeting all safety cornerstone objectives" to "unacceptable performance" (see Figure 19). The ROP recognizes that issues may range from very low to high safety significance, but plants are expected to address all issues effectively. The NRC performs very detailed baseline-level inspections at each plant. If plant problems arise, NRC oversight increases. The agency may perform supplemental inspections and take additional actions to ensure that significant performance issues are addressed. The latest plant-specific inspection findings and performance indicator information can be found on the NRC's Web site (see the Web Link Index).

The ROP takes into account improvements in the performance of the nuclear industry over the past 30 years and improved approaches to inspecting and evaluating the safety performance of NRC-licensed plants. The improvements in plant performance can be attributed both to successful regulatory oversight and to efforts within the nuclear industry. The ROP is described on the NRC's Web site and in NUREG-1649, Revision 4, "Reactor Oversight Process," issued December 2006.

Industry Performance Indicators

In addition to evaluating the performance of each individual plant, the NRC compiles data on overall reactor industry performance using various industry-level performance indicators (see Figure 20).

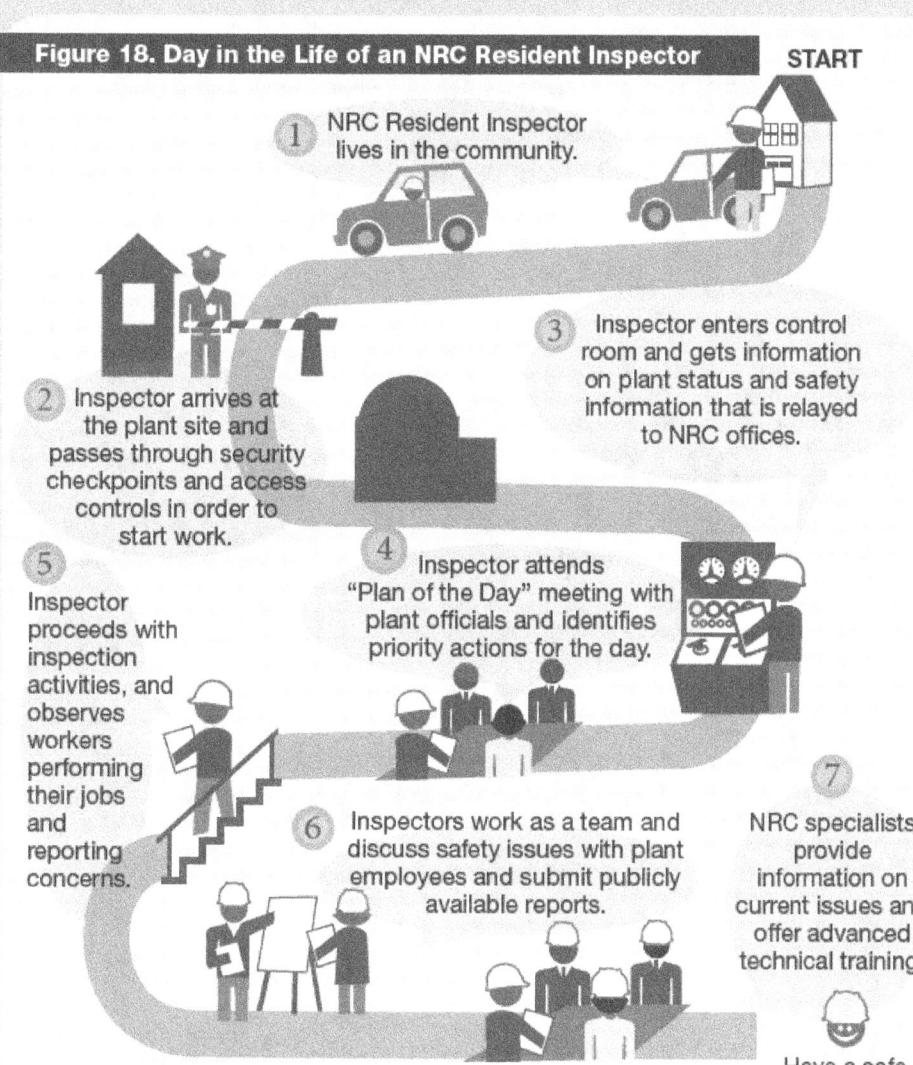

Figure 18. Day in the Life of an NRC Resident Inspector

START

① NRC Resident Inspector lives in the community.

③ Inspector enters control room and gets information on plant status and safety information that is relayed to NRC offices.

② Inspector arrives at the plant site and passes through security checkpoints and access controls in order to start work.

④ Inspector attends "Plan of the Day" meeting with plant officials and identifies priority actions for the day.

⑤ Inspector proceeds with inspection activities, and observes workers performing their jobs and reporting concerns.

⑥ Inspectors work as a team and discuss safety issues with plant employees and submit publicly available reports.

⑦ NRC specialists provide information on current issues and offer advanced technical training.

Have a safe day!

Figure 19. Reactor Oversight Action Matrix Performance Indicators

Performance Indicators

| GREEN | WHITE | YELLOW | RED |

Increasing Sa ety Signi icance →

Inspection Findings

| GREEN | WHITE | YELLOW | RED |

Increasing Sa ety Signi icance →

Figure 20. Industry Performance Indicators: FYs 2002–2011 Averages

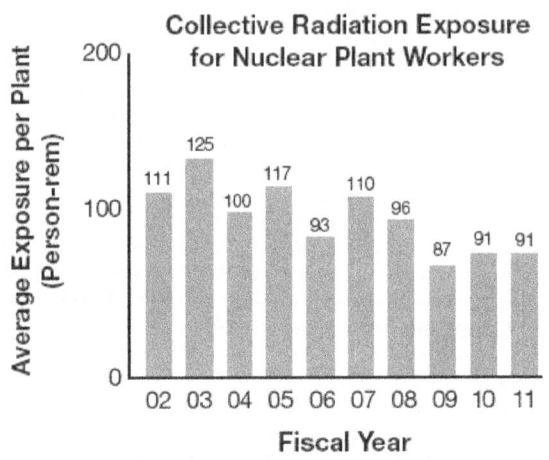

Collective Radiation Exposure for Nuclear Plant Workers

Average Exposure per Plant (Person-rem)

02: 111
03: 125
04: 100
05: 117
06: 93
07: 110
08: 96
09: 87
10: 91
11: 91

Fiscal Year

This indicator monitors the total radiation dose accumulated by plant personnel.

Further Explanation:
In 2011, those workers receiving a measurable dose of radiation received an average of about 0.1 rem. For comparison purposes, the average U.S. citizen receives 0.3 rem of radiation each year from natural sources (i.e., the everyday environment). See the definition of "exposure" in the Glossary.

Note: Data represent annual industry averages, with plants in extended shutdown excluded. Data are rounded for display purposes. These data may differ slightly from previously published data as a result of refinements in data quality.

Source: Licensee data as compiled by the NRC

Figure 20. Industry Performance Indicators: FYs 2002–2011 Averages (continued)

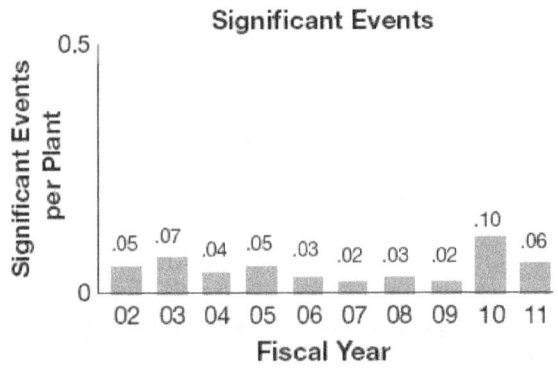

Significant Events

Significant Events per Plant

02: .05
03: .07
04: .04
05: .05
06: .03
07: .02
08: .03
09: .02
10: .10
11: .06

Fiscal Year

Significant events are events that meet specific NRC criteria, for example, degradation of safety equipment, a sudden reactor shutdown with complications, or an unexpected response to a sudden degradation of fuel or pressure boundaries. The NRC staff identifies significant events through detailed screening and evaluation of operating experience.

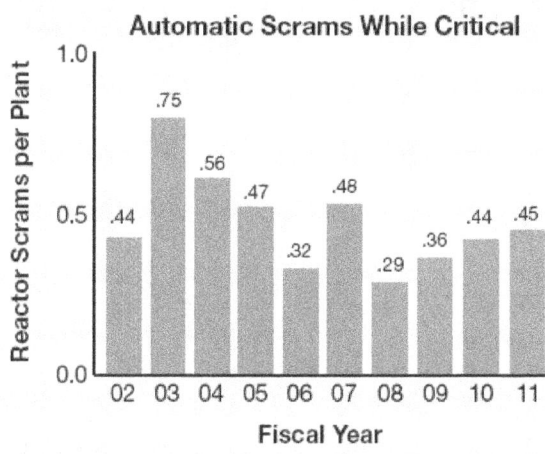

A reactor is said to be "critical" when it achieves a self-sustaining nuclear chain reaction, such as when the reactor is operating. The sudden shutting down of a nuclear reactor by the rapid insertion of control rods, either automatically or manually by the reactor operator, is referred to as a "scram." This indicator measures the number of unplanned automatic scrams that occurred while the reactor was critical.

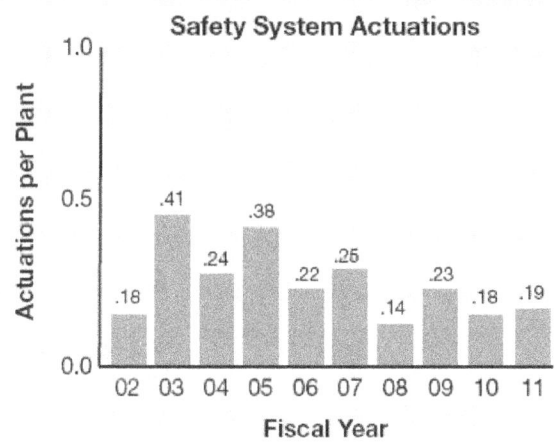

Safety system actuations are certain manual or automatic actions taken to start emergency core cooling systems or emergency power systems. These systems are specifically designed to either remove heat from the reactor fuel rods if the normal core cooling system fails or provide emergency electrical power if the normal electrical systems fail.

Reactor License Renewal

Based on the Atomic Energy Act of 1954, as amended, the NRC issues licenses for commercial power reactors to operate for 40 years. Under current regulations, licensees may renew their licenses for up to 20 years. Economic and antitrust considerations, not limitations of nuclear technology, determined the original 40-year term for reactor licenses. However, because of this selected time period, some systems, structures, and components may have been engineered on the basis of an expected 40-year service life.

As of June 2012, over 80 percent of the 104 licensed reactor units either have received or are under review for license renewal (31 units operate under their original license) . Of these, 73 units (at 44 sites) have received renewed licenses (see Figure 21). Figure 22 illustrates the years of commercial operation of operating power reactors. Figure 23 shows the expiration dates of operating commercial nuclear licenses. The decision to seek license renewal rests entirely with nuclear power plant owners and typically is based on the plant's economic situation and on whether it can meet NRC requirements.

See Appendix F and G for power reactors operating licenses by year issued and expired

The license renewal review process provides continued assurance that the current licensing basis will maintain an acceptable level of safety for the period of extended operation. The NRC will renew a license only if it determines that a currently operating plant will continue to maintain the required level of safety. Over the plant's life, this level of safety is enhanced through maintenance of the plant and its licensing basis, with appropriate adjustments to address new information from industry operating experience. The NRC regulations establish clear requirements for license renewal to ensure safe plant operation for extended plant life, as codified in 10 CFR Part 54, "Requirements for Renewal of Operating Licenses for Nuclear Power Plants." Environmental protection requirements for license renewal are contained in 10 CFR Part 51, "Environmental Protection Regulations for Domestic Licensing and Related Regulatory Functions."

The review of a renewal application proceeds along two paths—one for the review of safety issues and the other for environmental issues (see Figure 24). An applicant must provide the NRC with an evaluation that addresses the technical aspects of plant aging and describes the ways those effects will be managed. The applicant must also prepare for and evaluate the potential impact on the environment if the plant operates for up to an additional 20 years.

Figure 21. License Renewals Granted for Operating Nuclear Power Reactors

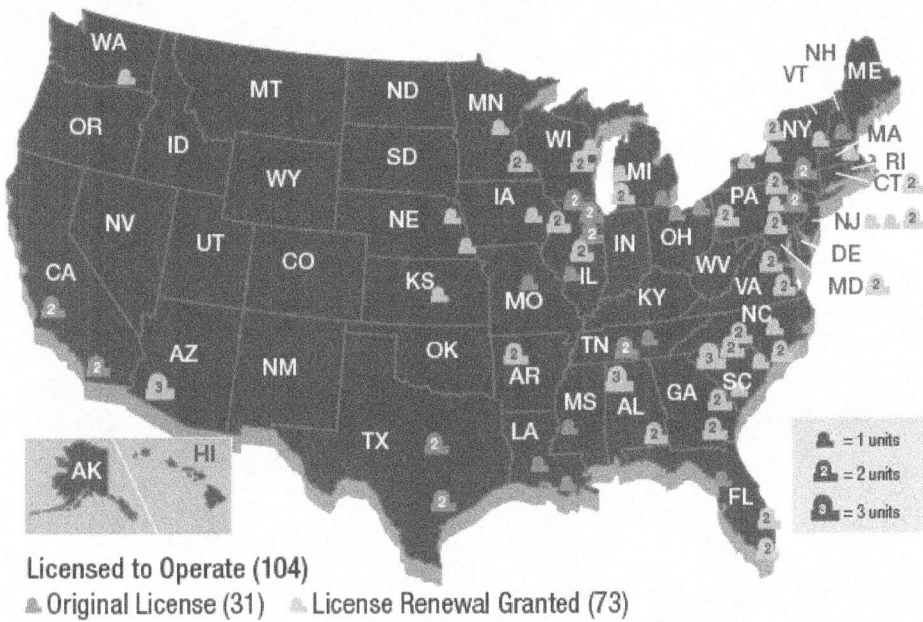

Licensed to Operate (104)
▲ Original License (31) ▲ License Renewal Granted (73)

Figure 22. U.S. Commercial Nuclear Power Reactors—Years of Operation by the End of 2012

2 reactors	37 reactors	50 reactors	15 reactors
10–19 years	20–29 years	30–39 years	>40 years

Note: Ages have been rounded up to the end of the year.

Figure 23. U.S. Commercial Nuclear Power Reactor Operating Licenses— Expiration by Year

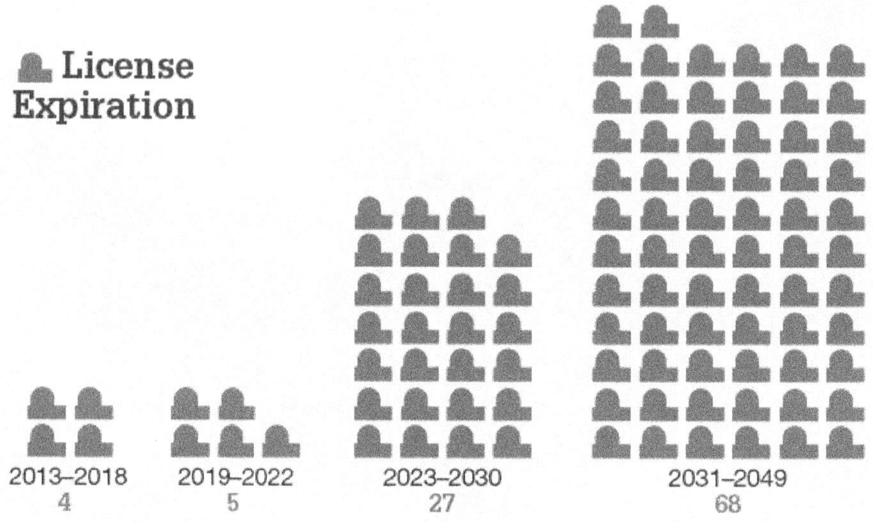

License Expiration

2013–2018	2019–2022	2023–2030	2031–2049
4	5	27	68

Figure 24. License Renewal Process

Opportunities for public interaction
★ If a request for a hearing is granted
★★ Available at www.nrc.gov

Safety Review
10 CFR Part 54

Onsite
Inspection(s)

 Safety
Evaluation
Audit & Review

Inspection
Reports
Issued**

License Renewal
Process and
Environmental
Scoping Meetings

 Safety Evaluation
Report with Open
Item(s) Issued**

Environmental
Review
10 CFR Part 51

Advisory
Committee on
Reactor Safeguards
(ACRS) Review

Site Environment Audit

 START

Safety
Evaluation
Reports
Issued**

ACRS
Review**

License
Renewal
Application**

 Draft
Supplement to
Generic
Environmental
Impact
Statement
(GEIS) Issued**

Hearings*

ACRS
Letter
Issued**

  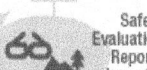 

Draft
Supplemental
Environmental
Impact
Statement
Public Meeting

Final
Supplement to
GEIS Issued**

NRC Decision
on Application**

Nuclear Reactors

42

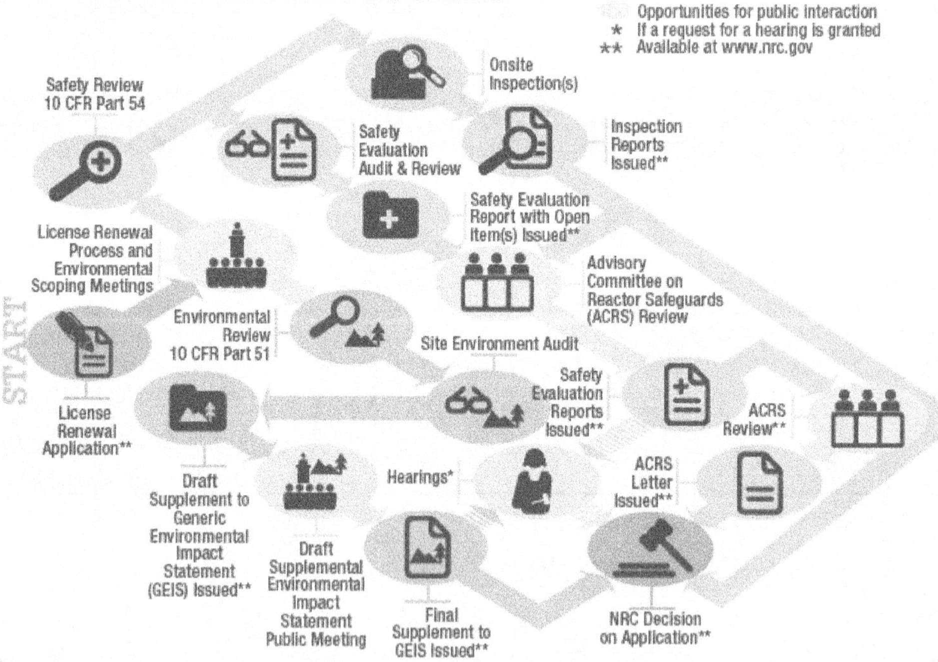

The NRC reviews the application and verifies the safety evaluation through onsite inspections.

Public Involvement

Public participation is an important part of the license renewal process. Members of the public have several opportunities to question how aging will be managed during the period of extended operation. The NRC makes available to the public information provided by the applicant and holds several public meetings. The agency fully documents its technical and environmental review results and makes them publicly available. In addition, ACRS holds public meetings to discuss technical or safety issues related to plant designs or a particular plant or site. Stakeholder concerns may be litigated in an adjudicatory hearing if any party that would be affected requests a hearing and submits an admissible contention. For more information, visit the NRC Web site (see the Web Link Index).

Research and Test Reactors

Nuclear research and test reactors (RTRs) are designed and used for research, testing, and education in nuclear engineering, physics, chemistry, biology, anthropology, medicine, materials sciences, and related fields. These reactors do not produce commercial electricity, but they help prepare people for nuclear-related careers in the fields of nuclear engineering, electric power, national defense, health services, research, and education.

See Appendix I for a list of the 31 operating RTRs regulated by the NRC and Appendix J for a list of the 11 RTRs regulated by the NRC that are in the process of decommissioning.

The largest U.S. RTR (at 20 megawatts thermal) (MWt) is 75 times smaller than the smallest U.S. commercial power nuclear reactor (at 1,500 MWt). There are 42 licensed RTRs:

- 31 RTRs operating in 21 States (see Figure 25)

- 11 RTRs shut down and in various stages of decommissioning

RTRs licensed to operate at a power level of 2 MWt or greater are inspected annually. RTRs licensed to operate at power levels below 2 MWt are inspected every 2 years. Since 1958, 83 licensed RTRs have been decommissioned.

Figure 25. U.S. Nuclear Research and Test Reactors

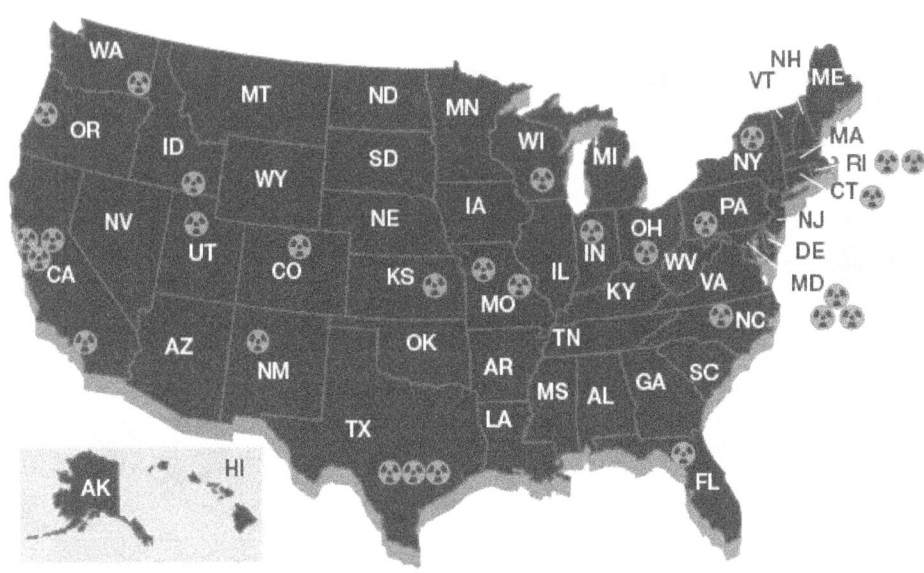

⚛ RTRs Licensed/Currently Operating (31)

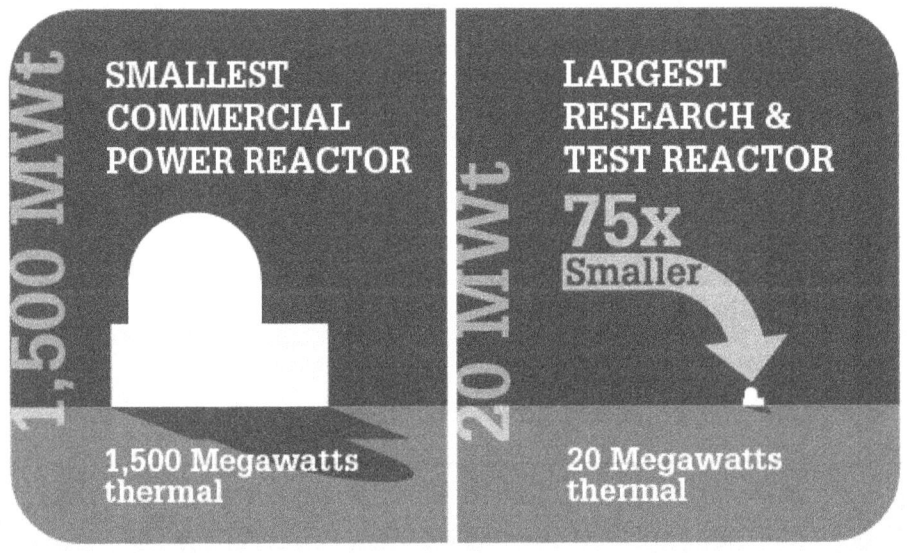

SMALLEST COMMERCIAL POWER REACTOR

1,500 MWt

1,500 Megawatts thermal

LARGEST RESEARCH & TEST REACTOR

20 MWt

75x Smaller

20 Megawatts thermal

Principal Licensing and Inspection Activities

The NRC's principal licensing and inspection activities related to RTRs include the following:

- licensing the 31 operating RTRs, including license renewals and license amendments;
- licensing approximately 100 RTR operators;
- requalifying operators' license before renewal; and
- conducting approximately 36 RTR inspections each year.

New Commercial Nuclear Power Reactor Licensing

The NRC is reviewing new reactor applications using a licensing process that substantially improved the system used through the 1990s (see Figure 26). In early 2012, the NRC issued the first combined construction and operating licenses (called a combined license or COL) under the new licensing process.

The NRC expects to review approximately 10 additional COL applications for approximately 16 new reactors over the next several years and has in place the infrastructure and staff to support the necessary technical work (see Figure 27 and the Web Link Index). The Fukushima lessons learned are being included in the design certification, COL, and ESP reviews.

Construction and Operating License Applications

As of June 2012, the NRC has received 18 COL applications for 28 new reactor units:

- Calvert Cliffs (MD)
- South Texas Project (TX)
- Bellefonte (AL)
- North Anna (VA)
- William States Lee III (SC)
- Shearon Harris (NC)
- Grand Gulf (MS)
- Vogtle (GA)*
- V.C. Summer (SC)*

- Callaway (MO)
- Levy County (FL)
- Victoria County Station (TX)
- Fermi (MI)
- Comanche Peak (TX)
- River Bend (LA)
- Nine Mile Point (NY)
- Bell Bend (PA)
- Turkey Point (FL)

* Approved by the NRC in early 2012

Figure 26. New Reactor Licensing Process

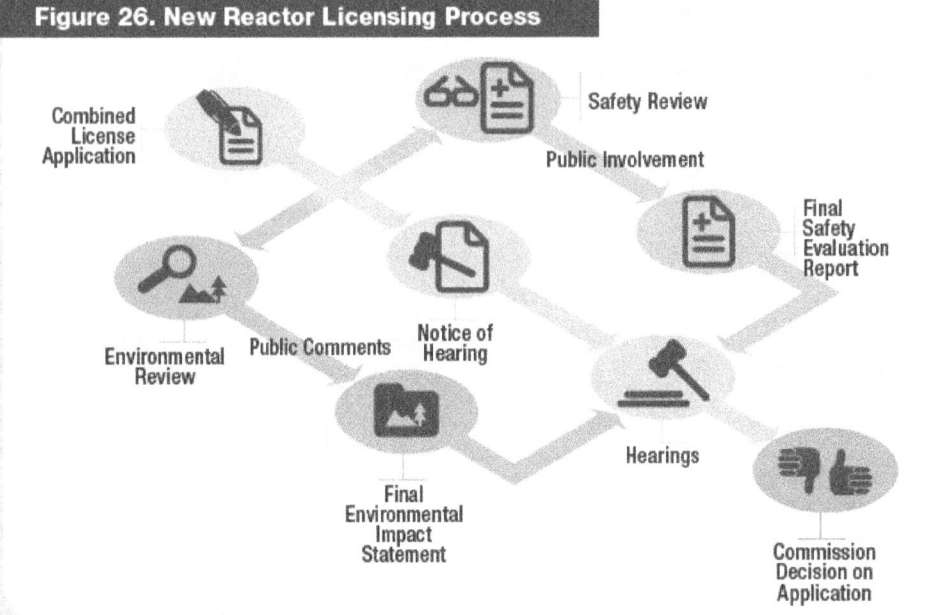

Figure 27. Locations of New Nuclear Power Reactors Applications

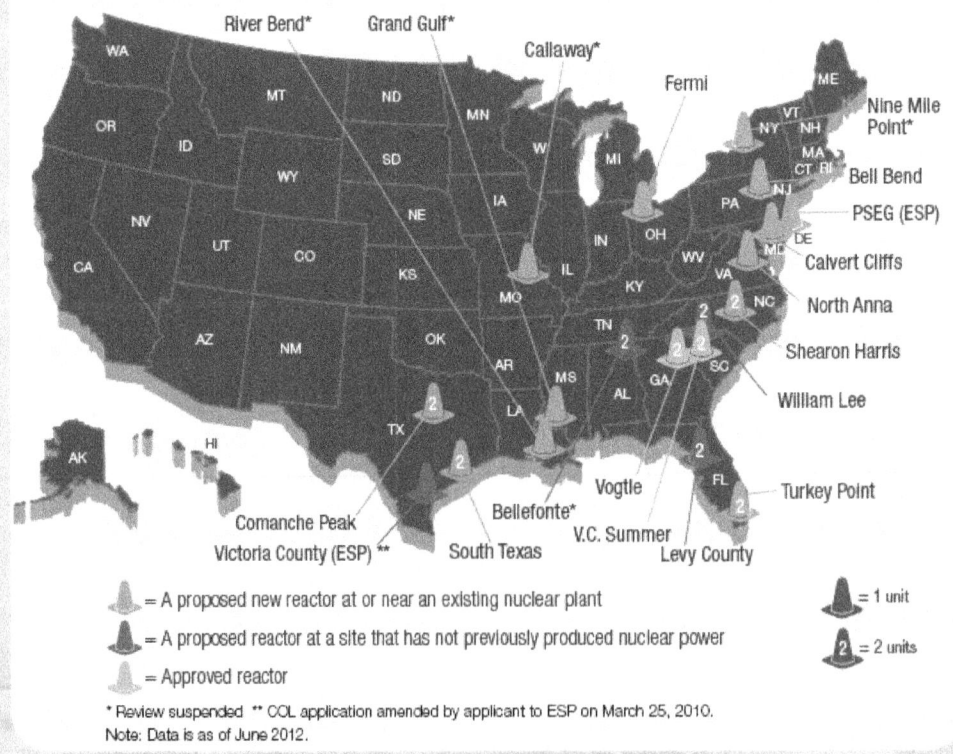

= A proposed new reactor at or near an existing nuclear plant

= A proposed reactor at a site that has not previously produced nuclear power

= Approved reactor

= 1 unit

= 2 units

* Review suspended ** COL application amended by applicant to ESP on March 25, 2010.
Note: Data is as of June 2012.

Nuclear Reactors

The NRC suspended or cancelled six COL application reviews because of changes in applicant business strategies (Grand Gulf, Callaway, Nine Mile Point, River Bend, Victoria County Station, and Bellefonte).

As of June 2012, the NRC had 10 COL applications for 16 units under active review. Figure 27 shows the locations of the potential new reactor sites. For the current review schedule for reactor licensing applications, consult the NRC public Web site (see the Web Link Index).

Public Involvement

The NRC's new reactor licensing process offers many opportunities for public participation. Before it receives an application, the agency uses public meetings to talk to residents in the community near the location where a proposed new reactor may be built to explain how the NRC reviews an application and how the public may participate in the process. Next, the NRC listens to comments on which factors should be considered in the agency's environmental review of the application. The public may then comment on the NRC's draft environmental evaluation that is posted on the agency's Web site. There is no formal opportunity for public comment on the staff's safety evaluation, but members of the public are welcome to attend public meetings and make comments. In addition, the public is afforded the opportunity to legally challenge a license application through Atomic Safety and Licensing Board hearings that are announced in press releases and posted on the NRC Web site. The NRC has tailored its new reactor licensing activities to review new applications effectively and efficiently without compromising safety.

Early Site Permits

An early site permit (ESP) provides for early resolution of site safety, environmental protection, and emergency preparedness issues independent of a specific nuclear plant review. The ACRS reviews those portions of the ESP application that concern safety. Mandatory adjudicatory hearings associated with the ESPs are conducted after the completion of the NRC staff's technical review.

The NRC has issued ESPs to the following applicants:

* System Energy Resources, Inc. (Entergy), for the Grand Gulf site in Mississippi

* Exelon Generation Company, LLC, for the Clinton site in Illinois

* Dominion Nuclear North Anna, LLC, for the North Anna site in Virginia

* Southern Nuclear Operating Company, for the Vogtle site in Georgia

Table 1. U.S. New Nuclear Power Plant Applications

Company (Project/Docket #)	Date of Application	Design	Date Accepted	Site Under Consideration	State	Existing Op. Plant
Calendar Year (CY) 2007 Applications						
NRG Energy (52-012/013)	9/20/07	ABWR	11/29/07	South Texas Project (2 units)	TX	Y
NuStart Energy (52-014/015)	10/30/07	AP1000	1/18/08	Bellefonte (2 units)	AL	N
UNISTAR (52-016)	7/13/07 (Env.), 3/13/08 (Safety)	EPR	1/25/08 6/3/08	Calvert Cliffs (1 unit)	MD	Y
Dominion (52-017)*	11/27/07	US-APWR	1/28/08	North Anna (1 unit)	VA	Y
Duke (52-018/019)	12/13/07	AP1000	2/25/08	William Lee Nuclear Station (2 units)	SC	N
2007 TOTAL NUMBER OF APPLICATIONS = 5 TOTAL NUMBER OF UNITS = 8						
CY 2008 Applications						
Progress Energy (52-022/023)	2/19/08	AP1000	4/17/08	Harris (2 units)	NC	Y
NuStart Energy (52-024)	2/27/08	ESBWR	4/17/08	Grand Gulf (1 unit)	MS	Y
Southern Nuclear Operating Co. (52-025/026)	3/31/08	AP1000	5/30/08	Vogtle (2 units)	GA	Y
South Carolina Electric & Gas (52-027/028)	3/31/08	AP1000	7/31/08	Summer (2 units)	SC	Y
Progress Energy (52-029/030)	7/30/08	AP1000	10/6/08	Levy County (2 units)	FL	N
Detroit Edison (52-033)	9/18/08	ESBWR	11/25/08	Fermi (1 unit)	MI	Y
Luminant Power (52-034/035)	9/19/08	US-APWR	12/2/08	Comanche Peak (2 units)	TX	Y
Entergy (52-036)	9/25/08	ESBWR	12/4/08	River Bend (1 unit)	LA	Y
AmerenUE (52-037)	7/24/08	EPR	12/12/08	Callaway (1 unit)	MO	Y
UNISTAR (52-038)	9/30/08	EPR	12/12/08	Nine Mile Point (1 unit)	NY	Y
PPL Generation (52-039)	10/10/08	EPR	12/19/08	Bell Bend (1 unit)	PA	Y
2008 TOTAL NUMBER OF APPLICATIONS = 11 TOTAL NUMBER OF UNITS = 16						
CY 2009 Applications						
Florida Power & Light Co. (52-040/041)	6/30/09	AP1000	9/4/09	Turkey Point (2 units)	FL	Y
2009 TOTAL NUMBER OF APPLICATIONS = 1 TOTAL NUMBER OF UNITS = 2						
CY 2010–2012 Applications						
No COL applications returned in CY 2010–2012.						
2010–2012 TOTAL NUMBER OF APPLICATIONS = 0 TOTAL NUMBER OF UNITS = 0						
CY 2013 Applications						
Blue Castle Project		TBD		Utah (1 unit)	UT	N
AmerenUE		TBD		Calloway (1 unit)	MO	Y
2013 TOTAL NUMBER OF APPLICATIONS = 2 TOTAL NUMBER OF UNITS = 2						
CY 2014 Applications						
One COL application is expected in fourth quarter of CY 2014.						
2014 TOTAL NUMBER OF APPLICATIONS = 1 TOTAL NUMBER OF UNITS = 6						
2007–2014 TOTAL NUMBER OF APPLICATIONS = 23 TOTAL NUMBER OF UNITS = 296						

☐ – Accepted/Docketed   ☒ – Expected   ☐ – Approved

* Design technology was changed by the applicant on June 28, 2010.

Note: Application updates in this table do not show all projects previously mentioned because of changes in intent status or conversion to an ESP from a COL application. Data are shown as of June 30, 2012.

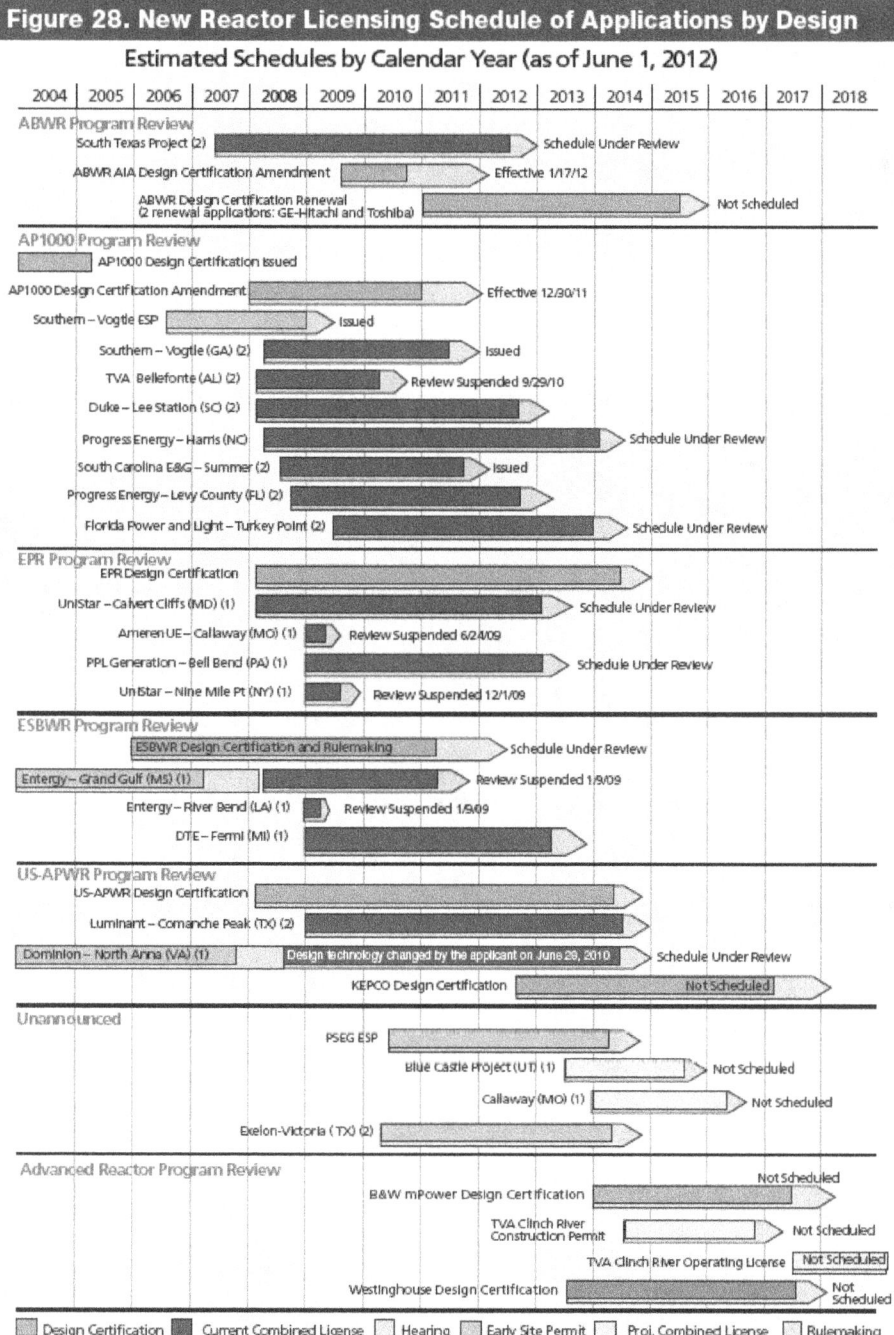

Figure 28. New Reactor Licensing Schedule of Applications by Design

Estimated Schedules by Calendar Year (as of June 1, 2012)

| 2004 | 2005 | 2006 | 2007 | 2008 | 2009 | 2010 | 2011 | 2012 | 2013 | 2014 | 2015 | 2016 | 2017 | 2018 |

ABWR Program Review
- South Texas Project (2) — Schedule Under Review
- ABWR AIA Design Certification Amendment — Effective 1/17/12
- ABWR Design Certification Renewal (2 renewal applications: GE-Hitachi and Toshiba) — Not Scheduled

AP1000 Program Review
- AP1000 Design Certification Issued
- AP1000 Design Certification Amendment — Effective 12/30/11
- Southern – Vogtle ESP — Issued
- Southern – Vogtle (GA) (2) — Issued
- TVA Bellefonte (AL) (2) — Review Suspended 9/29/10
- Duke – Lee Station (SC) (2)
- Progress Energy – Harris (NC) — Schedule Under Review
- South Carolina E&G – Summer (2) — Issued
- Progress Energy – Levy County (FL) (2)
- Florida Power and Light – Turkey Point (2) — Schedule Under Review

EPR Program Review
- EPR Design Certification
- UniStar – Calvert Cliffs (MD) (1) — Schedule Under Review
- Ameren UE – Callaway (MO) (1) — Review Suspended 6/24/09
- PPL Generation – Bell Bend (PA) (1) — Schedule Under Review
- UniStar – Nine Mile Pt (NY) (1) — Review Suspended 12/1/09

ESBWR Program Review
- ESBWR Design Certification and Rulemaking — Schedule Under Review
- Entergy – Grand Gulf (MS) (1) — Review Suspended 1/9/09
- Entergy – River Bend (LA) (1) — Review Suspended 1/9/09
- DTE – Fermi (MI) (1)

US-APWR Program Review
- US-APWR Design Certification
- Luminant – Comanche Peak (TX) (2)
- Dominion – North Anna (VA) (1) — Design technology changed by the applicant on June 28, 2010 — Schedule Under Review
- KEPCO Design Certification — Not Scheduled

Unannounced
- PSEG ESP
- Blue Castle Project (UT) (1) — Not Scheduled
- Callaway (MO) (1) — Not Scheduled
- Exelon-Victoria (TX) (2)

Advanced Reactor Program Review
- B&W mPower Design Certification — Not Scheduled
- TVA Clinch River Construction Permit — Not Scheduled
- TVA Clinch River Operating License — Not Scheduled
- Westinghouse Design Certification — Not Scheduled

Legend: Design Certification ▢ Current Combined License ▢ Hearing ▢ Early Site Permit ▢ Proj. Combined License ▢ Rulemaking

Note: Lines depict approximate dates on schedule. Data on projected applications are based on information from potential applicants and are subject to change. Schedules depicted for future activities represent nominal assumed review durations based on submittal timeframes in letters of intent from prospective applicants. Numbers in ( ) next to the COL name indicate the number of units per site. The acceptance review is included at the beginning of the COL review. The rules in 10 CFR Part 2, "Rules of Practice for Domestic Licensing Proceedings and Issuance of Orders," govern hearings on COLs.

On March 25, 2010, Exelon Nuclear Texas Holdings (Exelon) submitted an ESP application for the Victoria County Station site located in Victoria County, TX. Exelon previously submitted a COL application for the Victoria County Station site on September 2, 2008, and requested that the COL application be withdrawn when the NRC formally accepts the Victoria County Station ESP application.

On June 7, 2010, the NRC docketed the Victoria County ESP application. PSEG Power, LLC, and PSEG Nuclear, LLC (PSEG), submitted an ESP application in May 2010 on a site located near the Hope Creek/Salem site. The NRC expects to receive two additional ESP applications by the end of 2014.

Aerial view of Vogtle Units 3 and 4 construction site near Waynesboro, GA.

Design Certifications

The NRC has issued design certifications (DCs) for four reactor designs that can be referenced in an application for a nuclear power plant. A DC is valid for 15 years from the date of issuance, but it can be renewed for an additional 15 years. The new reactor designs incorporate new elements such as passive safety systems and simplified system designs.

These four designs are as follows:

- General Electric-Hitachi Nuclear Energy's (GEH's) Advanced Boiling-Water Reactor (ABWR)
- Westinghouse's System 80+
- Westinghouse's AP600
- Westinghouse's AP1000

The NRC is currently reviewing the following DC applications:

- AREVA's U.S. Evolutionary Power Reactor (EPR)
- Mitsubishi Heavy Industries' U.S. Advanced Pressurized-Water Reactor (US-APWR)

As of June 1, 2012, the NRC completed the technical reviews on GEH's Economic Simplified Boiling-Water Reactor (ESBWR).

In late 2011, the NRC completed rulemaking on Westinghouse's AP1000 DC amendment and STP Nuclear Operating Company's ABWR DC amendment to address the aircraft impact rule.

Design Certification Renewals

The NRC received two DC renewal applications for the ABWR from GEH and Toshiba in 2010. Renewals are good for 15 years.

Advanced Reactor Designs

A range of advanced reactor designs and technologies have emerged that may be submitted to the NRC within the next several years. These technologies include small-sized light-water reactors, liquid-metal reactors, and high-temperature gas-cooled reactors. The NRC will focus its advanced reactor efforts on ensuring that the agency is prepared to address the multiple new technologies being proposed. The NRC has been actively working to develop the regulatory framework in preparation for future licensing application submittals.

New Reactor Construction Inspections

The NRC established a special construction inspection organization in Region II in Atlanta, GA, to inspect licensee construction to ensure that it is performed in compliance with NRC-issued licenses and applicable regulations and to ensure that the as-built facility conforms to its COL. The NRC staff will examine the licensee's operational programs, such as security, radiation protection, and operator training and qualification, to ensure that the licensee is ready to operate the plant once it is built. The agency's construction site inspectors will verify a licensee's completion of inspections, tests, analyses, and acceptance criteria.

On February 10, 2012, the NRC issued a COL to Southern Nuclear Operating Company for Vogtle Units 3 and 4. On March 30, 2012, the NRC issued COLs to South Carolina Electric and Gas for V.C. Summer Units 2 and 3. The NRC provides oversight of the licensee and contractor activities under the Construction Reactor Oversight Process. This process periodically assesses licensee performance.

The NRC will use these direct inspections and other methods to confirm that the licensee has completed these actions and has met the acceptance criteria included in a COL before allowing startup of the plant.

The NRC has established resident inspector offices at both Vogtle and V.C. Summer. The inspectors will be at the site for the duration of the construction phase to oversee day-to-day activities of the licensee and its contractors. In addition, specialists in Region II's Center for Construction Inspection conduct periodic inspections at the site to ensure the facilities are being constructed in accordance with the approved design.

The agency also inspects vendor facilities to ensure that products and services furnished to new U.S. reactors meet quality and other regulatory requirements. The NRC has a vendor and quality assurance program and performs quality assurance inspections to ensure that licensees and their contractors meet the regulatory guidelines. To verify compliance with applicable regulations, the NRC inspects domestic and foreign vendors as well as the activities of applicants and licensees. More information on the NRC's new reactor licensing activities is available on the NRC Web site (see the Web Link Index).

Nuclear Regulatory Research

The NRC's research program supports the agency's regulatory mission by providing technical advice, tools, and information to identify and resolve safety issues, make regulatory decisions, and promulgate regulations and guidance. This includes conducting confirmatory experiments and analyses; developing technical bases that support the NRC's safety decisions; and preparing the agency for the future by evaluating the safety aspects of new technologies and designs for nuclear reactors, materials, waste, and security.

The research program focuses on challenges as the industry continues to evolve, including potential new safety issues, management of aging and material degradation issues, technical issues associated with the deployment of new technologies and reactor designs, and retention of technical skills as experienced staff retires. In the near term, research supports oversight of operating light-water reactors, the technology currently used in the United States. However, recent applications for advanced light-water reactors and preapplication activity regarding nonlight-water reactor vendors have prompted the agency to consider longer term research needs.

The NRC's research programs examine technical areas, such as:

- material degradation (e.g., stress-corrosion cracking, aging management, degradation mitigation technologies, boric acid corrosion, and embrittlement);

- new and evolving technologies (e.g., new reactor technology, mixed oxide fuel performance, digital instrumentation and control, and safety-critical software);

- experience gained from operating reactors;

- probabilistic risk assessment (PRA) methods;

- seismic and geotechnical hazards;

- ability of equipment to function in a harsh environment (e.g., heat, radiation, humidity);

- structural integrity assessments of reactor component degradation (e.g., nondestructive evaluation techniques and protocols); and

- human factors issues, including safety culture and computerization and automation of control rooms.

The research program also:

- Develops the agency's fire safety research programs, including fire modeling, fire PRA methods, and fire testing.

- Develops and improves computer codes as computational abilities expand and additional experimental and operational data allow for more realistic simulation. These computer codes analyze a wide spectrum of technical areas, including severe accidents, radionuclide transport through the environment, health effects of radioactive releases, nuclear criticality, fire conditions in nuclear facilities, thermal-hydraulic performance of reactors, reactor fuel performance, and nuclear power plant risk assessment.

- Ensures the secure use and management of nuclear facilities and radioactive materials by investigating potential security vulnerabilities and possible compensatory actions.

The NRC dedicates about 7 percent of its personnel and about 11 percent of its contracting funds to research. This research enables the NRC's highly skilled, experienced experts to formulate sound technical solutions based on science and to support timely and realistic regulatory decisions. The NRC research budget for FY 2012 is approximately $49.8 million. This includes contracts with national laboratories, universities, and other research organizations for greater expertise and access to research facilities. Figure 29 illustrates the primary areas of research. The NRC directs more than three-fourths of the research program toward maintaining the safety of existing operating reactors. The agency is also directing research in support of regulating new and advanced reactors. Radioactive waste programs and security are additional focus areas for research. Infrastructure support includes information technology and human resources. The NRC also has cooperative agreements with universities and nonprofit organizations to research specific areas of interest to the agency.

The NRC asked the National Academy of Sciences to assess the feasibility of doing a study on the cancer risk for populations around nuclear power facilities. The results of the scoping study (Phase 1 of the project) are publicly available and will be used to inform the epidemiological design of a potential Phase 2 cancer risk assessment.

See Appendix V for a list of cooperative agreements.

Figure 29. NRC Research Funding, FY 2012

Total $49.8 Million

- Reactor Program—$42.8 M
- New/Advanced Reactor Licensing—$3.7 M
- Homeland Security—$1.5 M
- Materials and Waste—$1.3 M
- Infrastructure Support—$0.4 M

Note: Totals may not equal sum of components because of rounding.

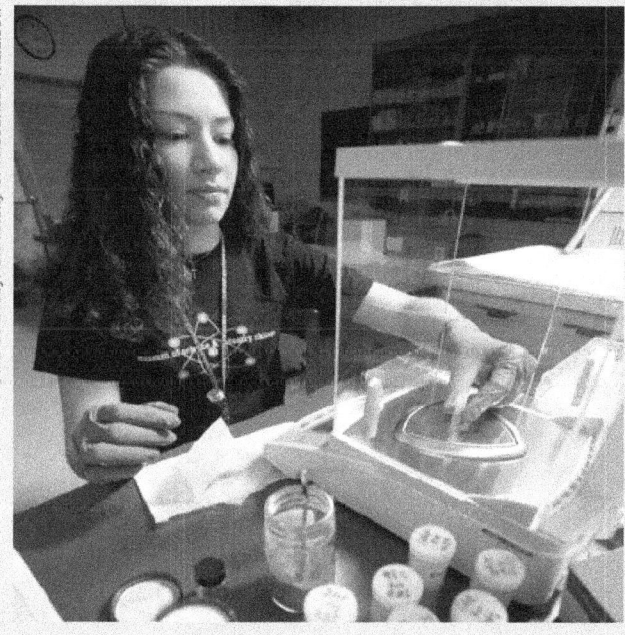

Photo courtesy: University of Wisconsin–Madison

Universities and other academic institutions use nuclear materials in laboratory experiments and to provide health physics support to other institutional nuclear materials users.

Over the last decades, significant advances have been made in the ability to assess seismic hazards for nuclear power plants in the United States. The NRC is currently sponsoring several projects in support of both an updated assessment of seismic hazard in the Central and Eastern United States (CEUS) and an enhancement of the overall framework under which the hazard characterizations are developed. The NRC, in collaboration with several other government agencies, recently issued a new seismic source characterization (SSC) model and report for use in seismic hazard assessments for nuclear facilities in the CEUS. The SSC model, developed over 3 years, replaces seismic source models developed in the late 1980s and can be used to calculate the likelihood of various levels of earthquake-caused ground motions. The new SSC model will be used by licensed nuclear power plants in the CEUS for seismic reevaluations, in addition to being used for licensing new nuclear facilities.

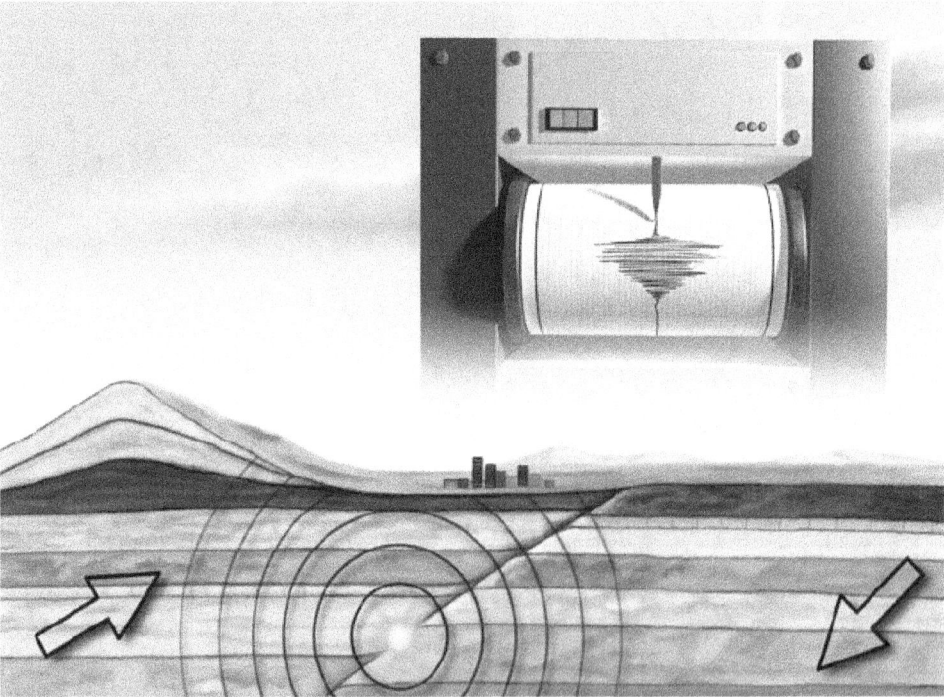

The NRC requires all of its licensees to take seismic activity into account when designing and maintaining its nuclear power plants. When new seismic hazard information becomes available, the NRC evaluates the new data and models and determines if any changes are needed at plants.

The State-of-the-Art Reactor Consequence Analyses (SOARCA) research project has developed best estimates of the offsite radiological health consequences for potential severe accidents for two U.S. nuclear power plants: the Peach Bottom Atomic Power Station, a BWR near Delta, PA, and the Surry Power Station, a PWR near Surry, VA. The project, which began in 2007, combined up-to-date information about the plants' layout and operations with local population data and emergency preparedness plans. This information was then analyzed using state-of-the-art computer codes that incorporate decades of research into severe reactor accidents. The draft report describing the Peach Bottom and Surry analyses, NUREG-1935, "State-of-the-Art Reactor Consequence Analyses (SOARCA) Report: Draft Report for Comment," issued January 2012, is publicly available. Upon conclusion of the project, the methods and models developed for the severe accident analyses in the SOARCA program will continue to be used to inform other agency programs.

The NRC collaborates with the international research community on both light-water and nonlight-water reactor technologies. This collaboration enables the agency to better leverage its resources, to initiate activities focused on evolutionary advances in existing technologies, and to determine the safety implications of new technologies. Collaboration is aided by the agency's leadership role in the standing committees and senior advisory groups of international organizations, such as IAEA and NEA.

The NRC also has research agreements with foreign governments for international cooperative research. The NRC is engaged in over 100 cooperative research agreements with more than two dozen countries and NEA, covering technical areas from severe accident research and code development to materials degradation, nondestructive examination, and human factors research. The agreements let the NRC leverage its own research expenditures by greatly reducing the cost of conducting research independently. They also afford the NRC access to facilities capable of research not currently possible in the United States.

Examples of agreements include.

- the NRC's Program to Assess Reliability of Emerging Nondestructive Techniques, with Finland, Japan, South Korea, Sweden, and Switzerland;

- more than 20 agreements with foreign regulators and research organizations for participation in the NRC's Cooperative Severe Accidents Research Program.

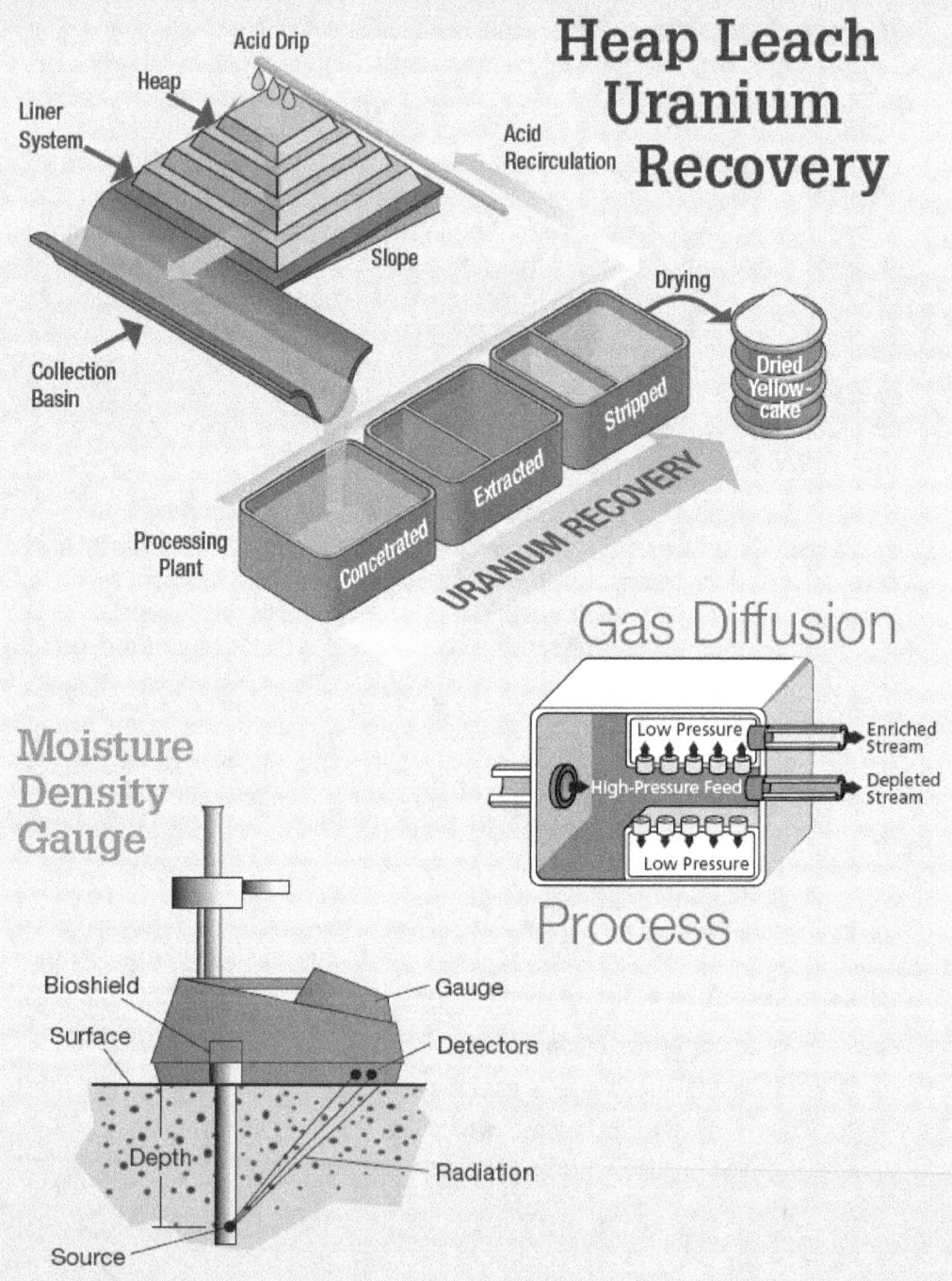

Heap Leach Uranium Recovery

Acid Drip

Heap

Liner System

Acid Recirculation

Slope

Collection Basin

Drying

Dried Yellow-cake

Processing Plant

Conceitrated

Extracted

Stripped

URANIUM RECOVERY

Gas Diffusion

Low Pressure

Enriched Stream

High-Pressure Feed

Depleted Stream

Low Pressure

Process

Moisture Density Gauge

Bioshield

Surface

Gauge

Detectors

Depth

Radiation

Source

Commercial Irradiator

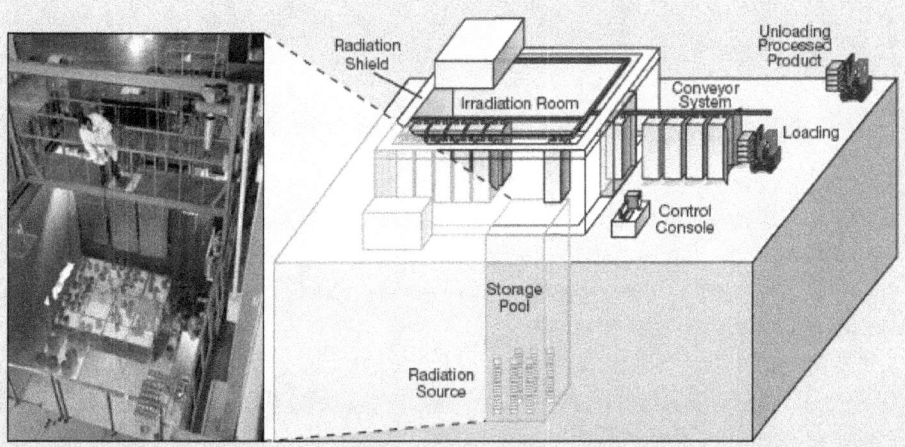

Nuclear Fuel Cycle

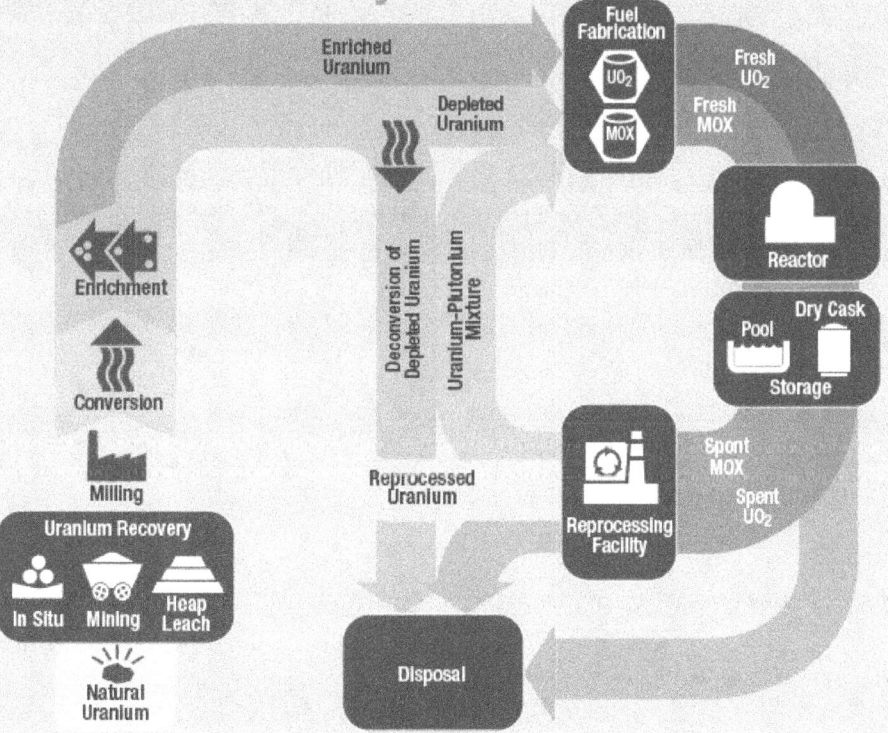

The NRC regulates nuclear materials for use in medical, industrial, and academic applications. It also regulates the phases of the nuclear fuel cycle, which begins with the uranium recovery, conversion, enrichment, and fabrication facilities that produce nuclear fuel for power plants.

Materials Licenses

Through agreements with the NRC, many States have assumed regulatory authority over radioactive materials, with the exception of nuclear reactors, fuel facilities, and certain quantities of special nuclear material. These States are called Agreement States. The NRC and Agreement State regulatory programs are designed to ensure that licensees use these materials safely and do not endanger public health and safety or cause damage to the environment.

See Appendix K for Agreement States

The NRC and Agreement States have issued approximately 21,800 licenses for general use of nuclear materials (see Table 2):

- The NRC administers approximately 2,900 licenses.

- 37 Agreement States administer approximately 18,900 licenses.

Reactor- and accelerator-produced radionuclides are used extensively throughout the United States for civilian and military industrial applications; basic and applied research; manufacture of consumer products; academic studies; and medical diagnosis, treatment, and research.

Medical and Academic

In both medical and academic settings, the NRC reviews the facilities, personnel, program controls, and equipment to ensure the safety of the public, patients, and workers who might be exposed to radiation.

Medical

The NRC and Agreement States issue licenses to hospitals and physicians for the use of radioactive materials in medical treatments. In addition, the NRC develops guidance and regulations for use by licensees and maintains a committee of medical experts to obtain advice about the use of radioactive materials in medicine.

Table 2. U.S. Materials Licenses by State

	Number of Licenses			Number of Licenses	
State	**NRC**	**Agreement States**	**State**	**NRC**	**Agreement States**
Alabama	18	439	Montana	89	0
Alaska	64	0	Nebraska	5	148
Arizona	12	366	Nevada	3	237
Arkansas	5	213	New Hampshire	8	82
California	57	1,852	New Jersey	39	638
Colorado	20	356	New Mexico	14	198
Connecticut	180	0	New York	22	1,403
Delaware	52	0	North Carolina	17	760
District of Columbia	42	0	North Dakota	8	83
Florida	22	1,720	Ohio	40	629
Georgia	17	520	Oklahoma	17	233
Hawaii	60	0	Oregon	5	335
Idaho	74	0	Pennsylvania	53	745
Illinois	32	711	Rhode Island	1	49
Indiana	283	0	South Carolina	15	414
Iowa	3	170	South Dakota	41	0
Kansas	11	286	Tennessee	22	589
Kentucky	9	431	Texas	49	1,665
Louisiana	11	519	Utah	10	197
Maine	2	125	Vermont	34	0
Maryland	84	598	Virginia	59	426
Massachusetts	25	500	Washington	15	405
Michigan	501	0	West Virginia	176	0
Minnesota	12	177	Wisconsin	14	321
Mississippi	6	331	Wyoming	84	0
Missouri	282	0	Others*	162	0
			Total	**2,886**	**18,871**

☐ Agreement State

* Others include major U.S. territories.

Note: The NRC and Agreement State data is as of June 2012.
The NRC licenses Federal agencies in Agreement States.

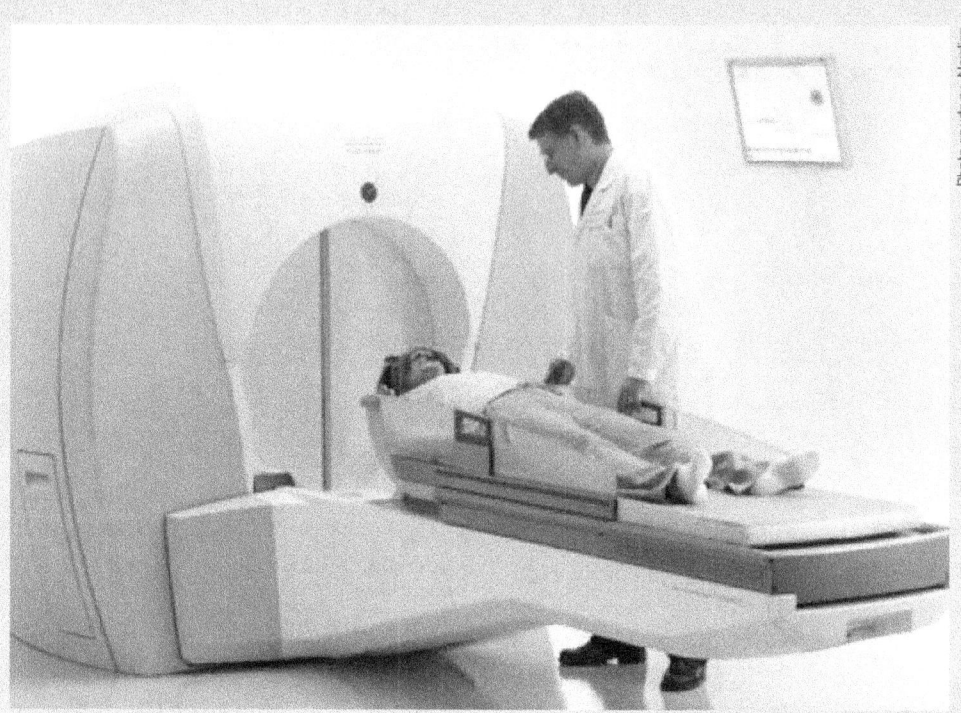

Photo courtesy: Nordion

Gamma Knife® used for treating brain tumors.

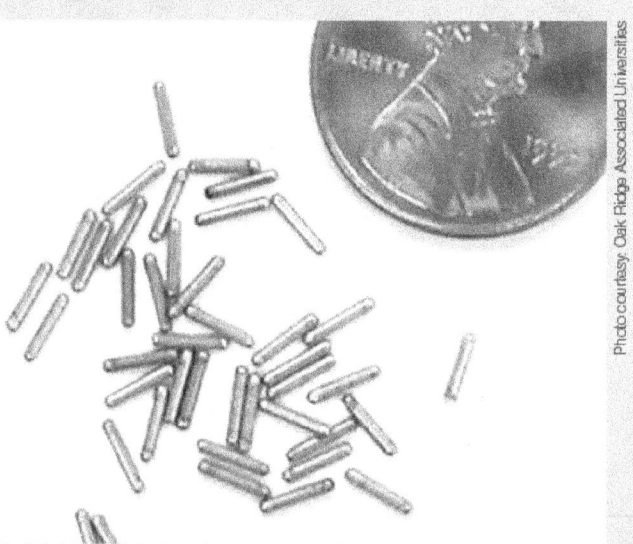

Photo courtesy: Oak Ridge Associated Universities

Iodine-125 and palladium-103 used in implantable seeds are primarily used to treat prostate cancer.

The NRC regulations require that physicians and physicists have special training and experience to practice radiation medicine. The training emphasizes safe operation of nuclear-related equipment and accurate recordkeeping. The Advisory Committee on the Medical Uses of Isotopes is comprised of physicians, scientists, and other health care professionals who advise the NRC staff on initiatives in the medical uses of radioactive materials.

Nuclear Medicine

About one-third of all patients admitted to hospitals are diagnosed or treated using radioactive materials. This branch of medicine is known as nuclear medicine, and the radioactive materials for treatment are called radiopharmaceuticals. Doctors of nuclear medicine use radiopharmaceuticals to diagnose patients through in vivo tests (direct administration of radiopharmaceuticals to patients) or in vitro tests (the addition of radioactive materials to lab samples taken from patients). Doctors also use radiopharmaceuticals and radiation-producing devices to treat conditions such as hyperthyroidism and certain forms of cancer and to ease pain caused by bone cancer. In the past decade, the use of nuclear medicine for treatment and diagnoses has increased significantly.

Diagnostic Procedures

For most diagnostic procedures in nuclear medicine, a small amount of radioactive material is administered, either by injection, inhalation, or by mouth. The radiopharmaceutical collects in the organ or area being evaluated, where it emits photons. These photons can be detected by a device known as a gamma camera, which produces images that provide information about the organ function and composition.

Radiation Therapy

The primary objective of radiation therapy is to deliver an accurately prescribed dose of radiation to the target site while minimizing the radiation dose to surrounding healthy tissue. Radiation therapy can be used to treat cancer or to relieve symptoms associated with certain diseases. Treatments often involve multiple exposures spaced over a period of time for maximum therapeutic effect. When used to treat malignant diseases, radiation therapy is often delivered in combination with surgery or chemotherapy.

There are three main categories of radiation therapy:

1. External beam therapy (also called teletherapy) is a beam of radiation directed to the target tissue. There are several different categories of external beam therapy units. The type of treatment machine that is regulated by the NRC contains a high-activity radioactive source (usually cobalt-60) that emits photons to treat the target site.

2. In brachytherapy treatments, sealed radioactive sources are permanently or temporarily placed near or on a body surface, in a body cavity, directly on a surface within a cavity, or directly on the cancerous tissue. The radiation dose is delivered at a distance of up to an inch (a few centimeters) from the target area.

3. Therapeutic radiopharmaceuticals are quantities of unsealed radioactive materials that localize in a specific region or organ system to deliver a large radiation dose.

Academic

The NRC issues licenses to academic institutions for educational and research purposes. For example, qualified instructors use radioactive materials in classroom demonstrations. Scientists in a wide variety of disciplines use radioactive materials for laboratory research.

Industrial

The NRC and Agreement States license users of radioactive material for the specific type, quantity, and location of material that may be used. Radionuclides are used in industrial and commercial applications, including industrial radiography, gauges, well logging, and manufacturing. For example, radiography uses radiation sources to find structural defects in metallic materials and welds. Gas chromatography uses low-energy radiation sources for identifying the chemical elements in an unknown substance. Gas chromatography can determine the components of complex mixtures, such as petroleum products, smog, and cigarette smoke, and can be used in biological and medical research to identify the components of complex proteins and enzymes. Well-logging devices use a radioactive source and detection equipment to make a record of geological formations down a bore hole. This process is used extensively for oil, gas, coal, and mineral exploration.

Nuclear Gauges

Nuclear gauges are used as nondestructive devices to measure the physical properties of products and industrial processes as a part of quality control. Gauges use radiation sources to determine the thickness of paper products, fluid levels in oil and chemical tanks, and the moisture and density of soils and material at construction sites. There are fixed and portable gauges.

A fixed gauge consists of a radioactive source that is contained in a source holder. When the user opens the container's shutter, a controlled beam of radiation hits the material or product being processed or controlled. A detector mounted opposite the source measures the radiation passing through the product. The gauge readout or computer monitor shows the measurement. The material and process being monitored dictate the selection of the type, energy, and strength of radiation.

Fixed fluid gauges are installed on a pipe that is used by the beverage, food, plastics, and chemical industries to measure the densities, flow rates, levels, thicknesses, and weights of a wide variety of materials and surfaces.

Figure 30 shows a portable gauge configuration in which the gamma source is placed under the surface of the ground through a tube. Radiation is then transmitted directly to the detector on the bottom of the gauge, allowing accurate measurements of compaction. Industry uses such gauges to monitor the structural integrity of roads, buildings, and bridges and to explore for oil, gas, and minerals. Airport security uses nuclear gauges to detect explosives in luggage at airports.

A portable gauge is a radioactive source and detector mounted together in a portable shielded device. The device is placed on the object to be measured, and the source is either inserted into the object or the gauge relies on a reflection of radiation from the source to bounce back to the bottom of the gauge. The detector in the gauge measures the radiation either directly from the inserted source or from the reflected radiation.

The radiation measurement indicates the thickness, density, moisture content, or some other property that is displayed on a gauge readout or on a computer monitor. The top of the gauge has sufficient shielding to protect the operator while the source is exposed. When the measuring process is completed, the source is retracted or a shutter closes, minimizing exposure from the source.

Figure 30. Moisture Density Gauge

Direct Transmission

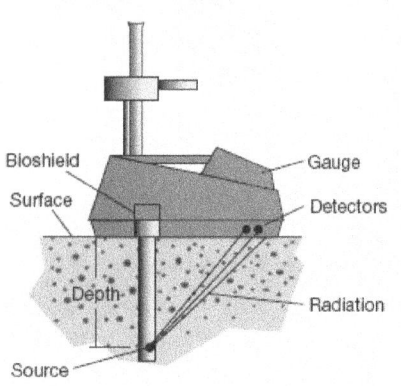

A moisture density gauge indicates whether a foundation is suitable for constructing a building or roadway.

Figure 31. Commercial Irradiator

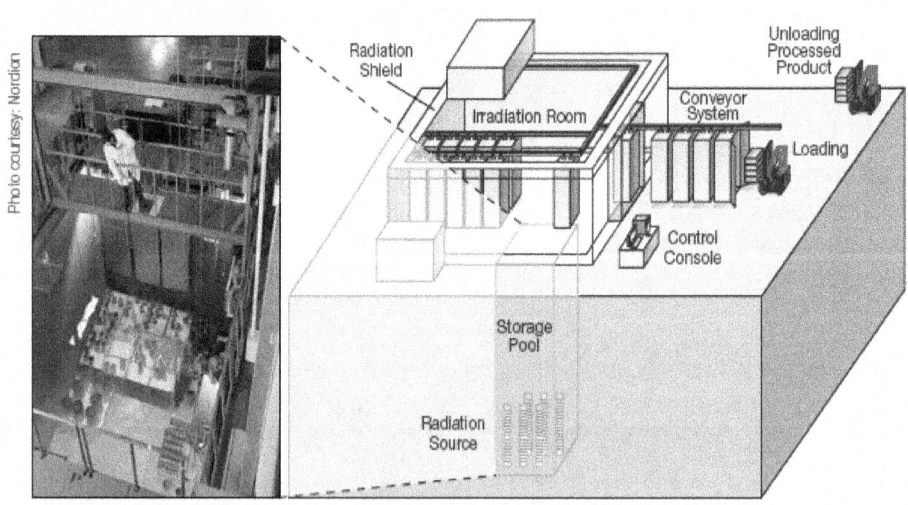

Nuclear Materials

Commercial Irradiators

Commercial irradiators expose products such as food, food containers, spices, medical supplies, and wood flooring to radiation to eliminate harmful bacteria, germs, and insects or for hardening or other purposes (see Figure 31). The gamma radiation does not leave any radioactive residue or cause any of the treated products to become radioactive themselves. The source of that radiation can be radioactive materials (e.g., cobalt-60), an x-ray tube, or an electron beam.

The NRC and Agreement States license approximately 50 commercial irradiators nationwide. For the past 40 years, the U.S. Food and Drug Administration and other agencies have approved the irradiation of meat and poultry, as well as other foods, including fresh fruits, vegetables, and spices. The amount of radioactive material in the devices can range from 1 curie to 10 million curies. NRC regulations protect workers and the public from radiation involved in irradiation operations. Generally, two types of commercial irradiators are in operation in the United States: underwater and wet-source-storage panoramic models.

In the case of underwater irradiators, the sealed sources (radioactive material encased inside a capsule) that provide the radiation remain in the water at all times, providing shielding for workers and the public. The product to be irradiated is placed in a watertight container, lowered into the pool, irradiated, and then removed.

With wet-source-storage panoramic irradiators, the radioactive sealed sources are also stored in the water, but they are raised into the air to irradiate products that are automatically moved in and out of the room on a conveyor system. Sources are then lowered back to the bottom of the pool. For this type of irradiator, thick concrete walls or steel barriers protect workers and the public when the sources are lifted from the pool.

Transportation

About 3 million packages of radioactive materials are shipped each year in the United States, either by road, rail, air, or water. This represents less than 1 percent of the Nation's yearly hazardous material shipments. Regulating the safety of commercial radioactive material shipments is the joint responsibility of the NRC and the U.S. Department of Transportation (DOT). The vast majority of these shipments consist of small amounts of radioactive materials used in industry, research, and medicine. The NRC requires such materials to be shipped in accordance with DOT's hazardous materials transportation safety regulations.

Material Security

In January 2009, the NRC deployed its National Source Tracking System (NSTS), by which the agency and the Agreement States track the manufacture, distribution, and ownership of the most high-risk sources. Licensees use the NSTS, a secure Web-based system, to enter up-to-date information on the receipt or transfer of tracked radioactive sources (see Figure 32). Over the past several years, the NRC and the Agreement States have increased the controls they have imposed on the most sensitive radioactive materials, including physical security requirements and limited personnel access to the materials. Working with other Federal agencies, such as DHS and the National Nuclear Security Administration, the NRC has also implemented a voluntary program of additional security improvements. Together, these activities help make potentially dangerous radioactive sources even more secure and less vulnerable to terrorists.

Principal Licensing and Inspection Activities

Each year, the NRC issues approximately 2,900 new licenses, license renewals, or amendments for existing material licenses. The NRC conducts approximately 1,250 health and safety and security inspections of its nuclear materials licensees each year.

Nuclear Fuel Cycle

Figure 33 illustrates the nuclear fuel cycle, which begins with the uranium recovery and continues with conversion, enrichment, and fabrication facilities that produce nuclear fuel for power plants. To make fuel for reactors, uranium is recovered or extracted from the ore, converted, enriched, and yellowcake is manufactured into fuel pellets.

Uranium Recovery

The NRC does not regulate traditional mining, but it does regulate the processing of uranium ore. It has jurisdiction over uranium recovery facilities such as conventional mills, heap leach facilities, and in situ recovery (ISR) facilities. The NRC has a well-established regulatory framework for ensuring that uranium recovery facilities are appropriately licensed, operated, decommissioned, and monitored to protect public health and safety.

Conventional Uranium Mill

A conventional uranium mill is a chemical plant that extracts uranium from mined ore. Conventional mills are typically located in areas of low population density, within about 50 kilometers (30 miles) of a uranium mine.

Figure 32. Life-Cycle Approach to Source Security

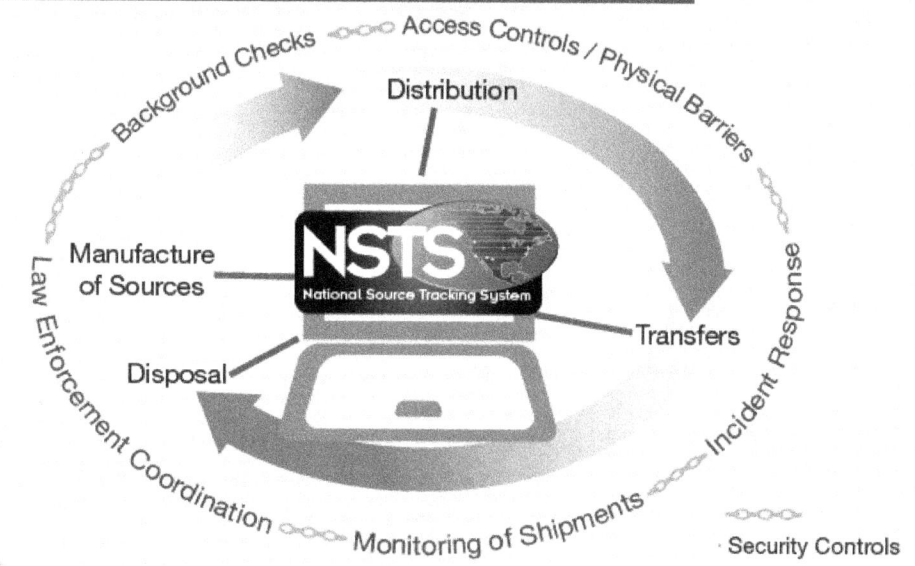

Figure 33. The Nuclear Fuel Cycle

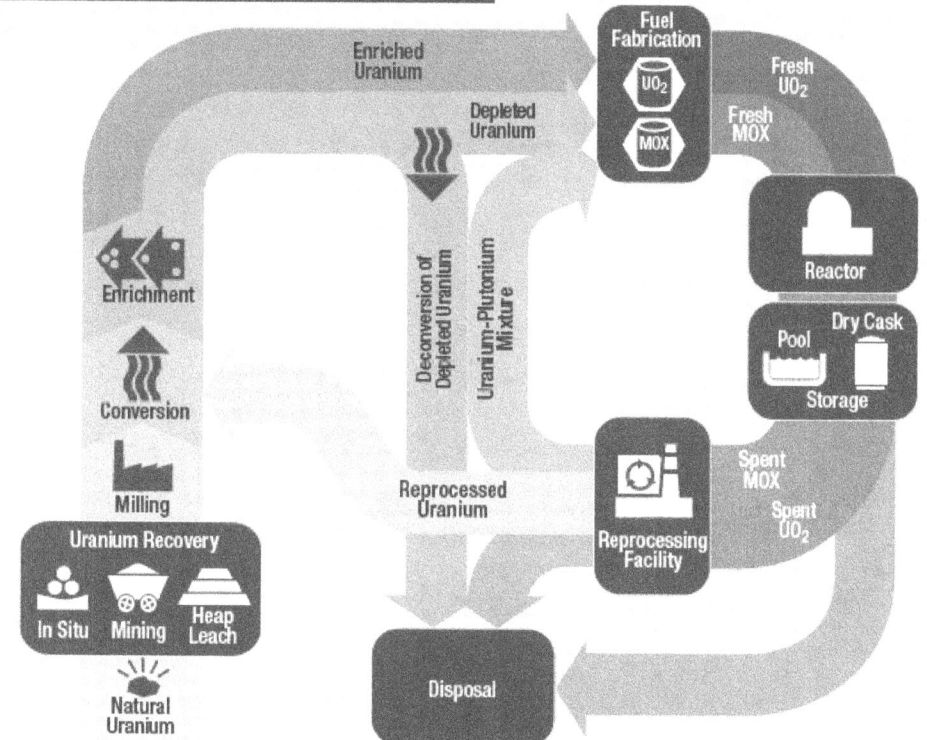

The mined ore is transported to the mill, where it is crushed. Sulfuric acid then dissolves the soluble components, including 90 to 95 percent of the uranium, from the ore. The uranium is then separated from the solution, concentrated, and dried to form yellowcake. There are 18 uranium recovery sites licensed by the NRC—11 are conventional mills and seven are ISR facilities. Of these, 10 are in various stages of decommissioning and one is in standby status with the potential to restart in the future.

Heap Leach Facility

Heap leach facilities are used to extract uranium. Uranium ore is transported to the site and placed in piles or heaps. These heaps are lined to prevent uranium and other chemicals from migrating into the subsurface. Sulfuric acid is dripped onto the heap, which dissolves uranium as it migrates through the ore. Uranium solution collects at the bottom of the heap and drains to collection basins, where it is piped to the processing plant. At the plant, uranium is concentrated, extracted, stripped, and dried to form yellowcake (see Figure 34). The NRC does not currently license any heap leach facilities; however, applications for such facilities are expected within the next few years.

In Situ Recovery

ISR is another means of extracting uranium—this time from underground ore. ISR facilities recover uranium from ores for which recovery may not be economically viable by other methods. In this process, a solution of native ground water typically mixed with oxygen or hydrogen peroxide and sodium bicarbonate or carbon dioxide is injected through wells into the ore to dissolve the uranium. The resulting solution is pumped from the rock formation, and the uranium is then separated from the solution to form yellowcake (see Figure 35). The United States has about 14 ISR facilities. Of these facilities, the NRC licenses seven and Agreement States license the rest (see Figure 36 and Table 3).

Because of the resurgence of interest in the construction of new nuclear power plants, the agency anticipates as many as 28 applications for new uranium recovery facilities and expansions or restarts of existing facilities in the next few years.

As of April 2012, the agency had received six applications for new facilities and six applications to expand or restart an existing facility. One new facility and one expansion application have been withdrawn; however, these applications may be resubmitted in the future.

Figure 34. The Heap Leach Recovery Process

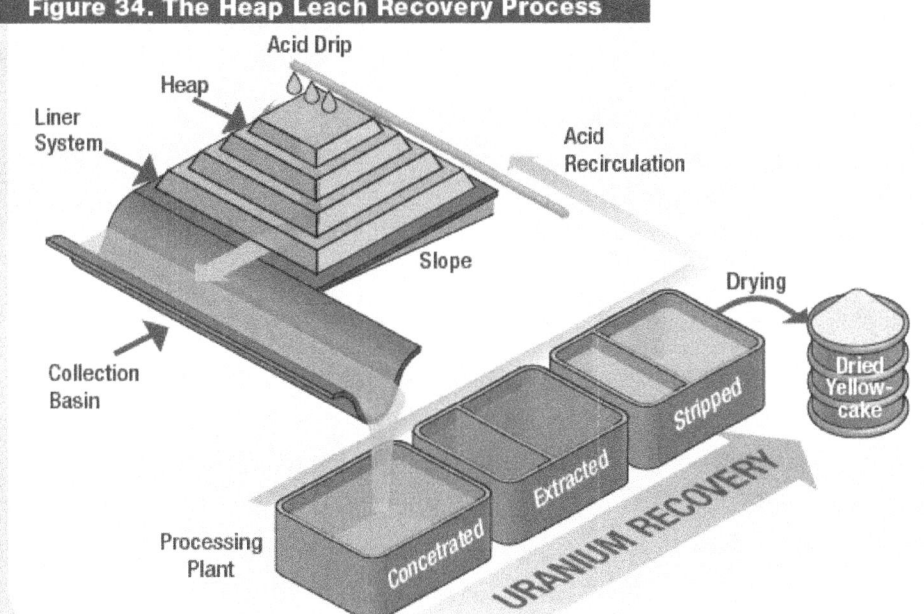

Acid Drip

Heap

Liner System

Acid Recirculation

Slope

Drying

Dried Yellow-cake

Collection Basin

Processing Plant

Concetrated

Extracted

Stripped

URANIUM RECOVERY

Figure 35. The In Situ Uranium Recovery Process

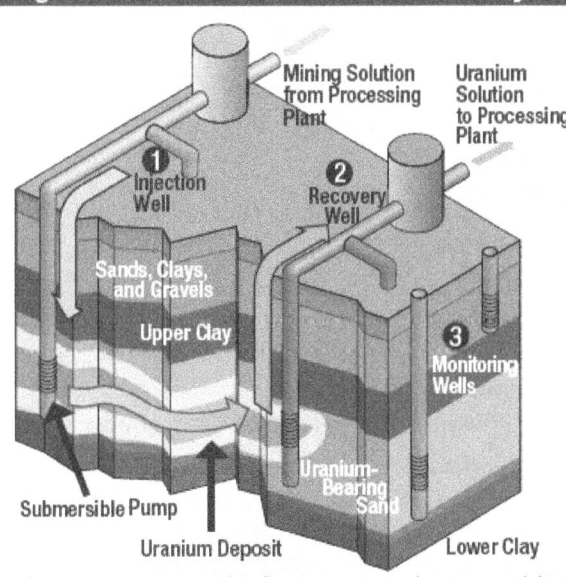

Mining Solution from Processing Plant

Uranium Solution to Processing Plant

❶ Injection Well

❷ Recovery Well

Sands, Clays, and Gravels

Upper Clay

❸ Monitoring Wells

Submersible Pump

Uranium-Bearing Sand

Uranium Deposit

Lower Clay

Injection wells (1) pump a chemical solution—typically groundwater mixed with sodium bicarbonate, hydrogen peroxide, and oxygen—into the layer of earth containing uranium ore. The solution dissolves the uranium from the deposit in the ground and is then pumped back to the surface through recovery wells (2) and sent to the processing plant to be processed into uranium yellowcake. Monitoring wells (3) are checked regularly to ensure that uranium and chemicals are not escaping from the drilling area.

The current status of applications can be found on the NRC's Web site (see the Web Link Index). Existing facilities and new potential sites are located in Wyoming, New Mexico, Nebraska, South Dakota, Oregon, and Nevada, and in the Agreement States of Texas, Colorado, and Utah (see Figure 37). The NRC works closely with stakeholders, including Native American Tribal governments, to address concerns with the licensing of new uranium recovery facilities. The NRC is also responsible for the following:

See Appendix M for Major
U.S. Fuel Cycle Facility Sites

- inspecting and overseeing both active and inactive uranium recovery facilities;

- ensuring that siting and design features of mill tailings (waste) minimize the release of radon and the disturbance of tailings by natural forces (see Glossary);

- developing requirements to ensure cleanup of active and formerly active uranium recovery facilities;

- formulating stringent financial requirements to ensure funds are available for decommissioning;

- monitoring adherence to requirements for below-grade disposal of mill tailings and liners for tailings impoundments;

- monitoring to prevent ground water contamination; and

- long-term monitoring and oversight of decommissioned facilities.

Fuel Cycle Facilities

The NRC licenses and routinely conducts safety, safeguards, and environmental protection inspections at all commercial fuel cycle facilities involved in conversion, enrichment, and fuel fabrication (see Figures 37–39). These special fuel facilities use a process that turns uranium from the ground into fuel for nuclear reactors. This process converts uranium yellowcake into uranium hexafluoride (UF_6), enriches the uranium in the isotope uranium-235, and fabricates ceramic fuel pellets. Fabrication is the final step in the process used to produce uranium fuel. Fabrication begins with the conversion of enriched UF_6 gas to a uranium dioxide (UO_2) solid. On average, the NRC completes approximately 85 new licenses, license renewals, license amendments, and safety and safeguards reviews for fuel cycle facilities annually. Fuel fabrication facilities mechanically and chemically process the enriched uranium into nuclear reactor fuel.

Figure 36. Locations of NRC-Licensed Uranium Recovery Facility Sites

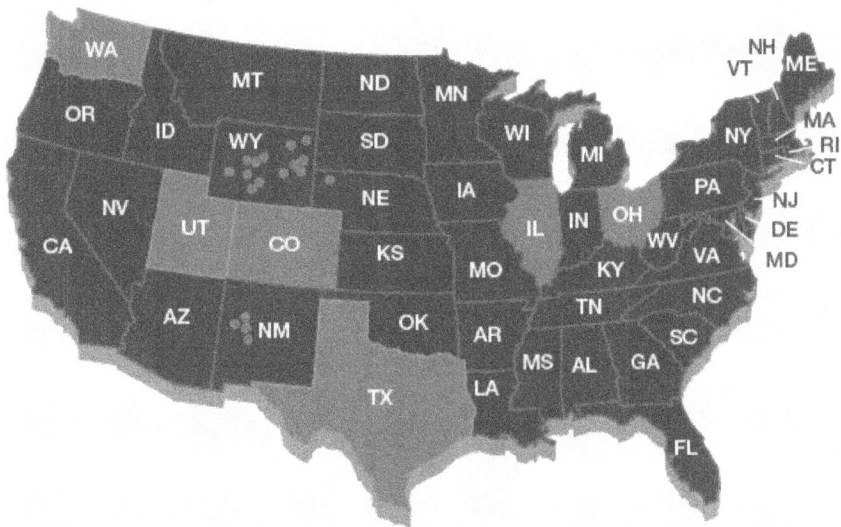

■ States with authority to license uranium recovery facility sites
■ States where the NRC has retained authority to license uranium recovery facilities
• NRC-licensed uranium recovery facility sites (18)

Table 3. Locations of NRC-Licensed Uranium Recovery Facilities

Licensee	Site Name, Location
In Situ Recovery Facilities	
Uranium One	Willow Creek, WY
Cameco Resources, Inc.	Crow Butte, NE*
Hydro Resources, Inc.°	Crownpoint, NM
Cameco Resources, Inc.	Smith Ranch and Highlands, WY*
Uranium One	Moore Ranch, WY
Lost Creek ISR, Inc.	Lost Creek, WY
Uranerz Energy Corp.	Nichols Ranch, WY
Conventional Uranium Mill Recovery Facilities	
American Nuclear Corp.†	Gas Hills, WY
Bear Creek Uranium Co.†	Bear Creek, WY
Exxon Mobil Corp.†	Highlands, WY
Homestake Mining Co.†	Homestake, NM
Kennecott Uranium Corp.°	Sweetwater, WY
Pathfinder Mines Corp.†	Lucky Mc, WY
Pathfinder Mines Corp.†	Shirley Basin, WY
Rio Algom Mining, LLC†	Ambrosia Lake, NM
Umetco Minerals Corp.†	Gas Hills, WY
United Nuclear Corp.†	Church Rock, NM
Western Nuclear, Inc.†	Split Rock, WY

Note: For further details on NRC-related uranium recovery facility applications in review and applications, restarts, and expansions, see the Web Link Index. This table does not include uranium recovery facilities licensed by Agreement States.

* Satellite facilities are located within the State.

† These sites are undergoing decommissioning.

° Hydro has an operating license, but the facility has not yet been constructed. Kennecott has an operating license but is in "standby" mode.

Figure 37. Locations of Fuel Cycle Facilities

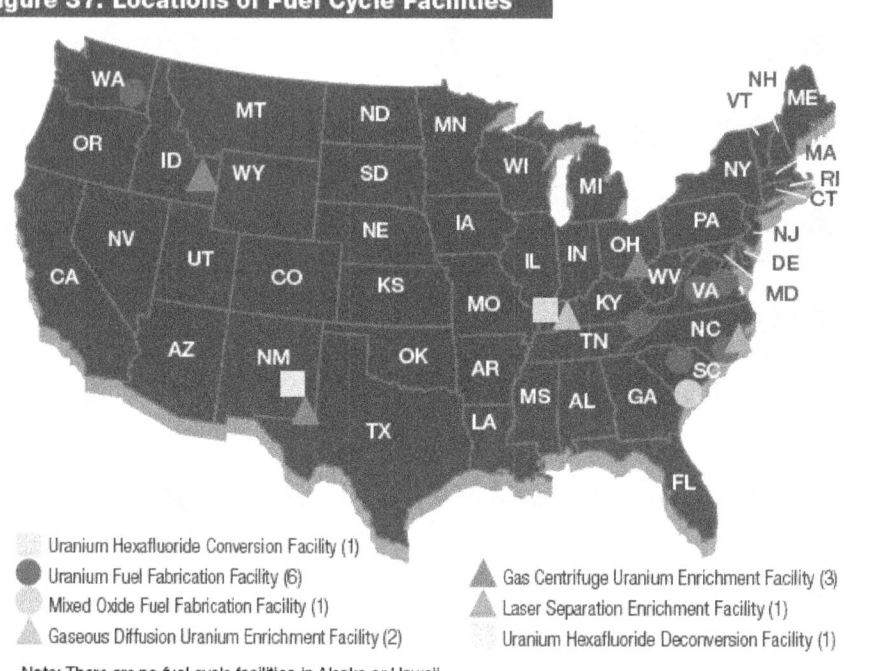

Uranium Hexafluoride Conversion Facility (1)
Uranium Fuel Fabrication Facility (6)
Mixed Oxide Fuel Fabrication Facility (1)
Gaseous Diffusion Uranium Enrichment Facility (2)

Gas Centrifuge Uranium Enrichment Facility (3)
Laser Separation Enrichment Facility (1)
Uranium Hexafluoride Deconversion Facility (1)

Note: There are no fuel cycle facilities in Alaska or Hawaii.

URENCO USA gas centrifuge uranium enrichment facility in Eunice, NM.

Figure 38. Enrichment Processes

A. Gaseous Diffusion Process

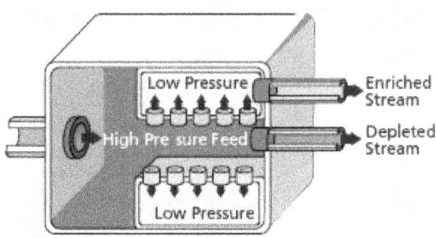

B. Gas Centrifuge Process

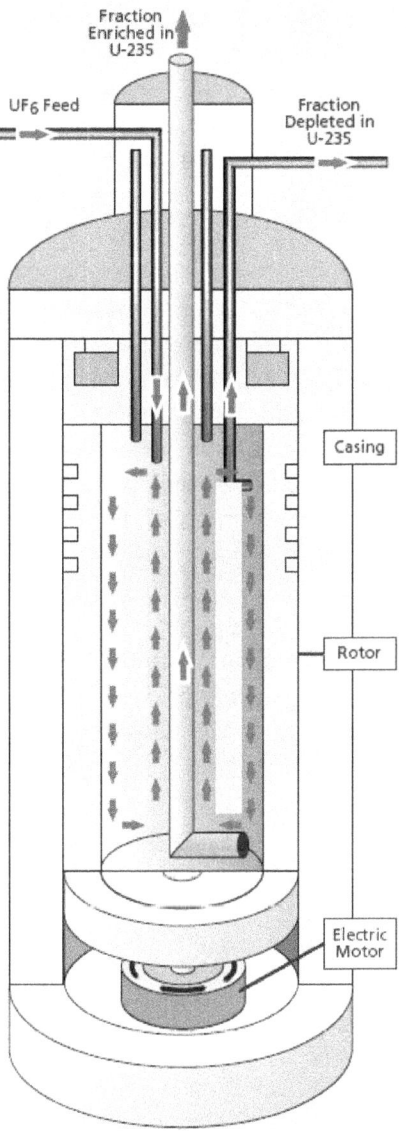

A. The gaseous diffusion process uses molecular diffusion to separate a gas from a two-gas mixture. The isotopic separation is accomplished by diffusing uranium, which has been combined with fluorine to form UF_6 gas, through a porous membrane (barrier) and using the different molecular velocities of the two isotopes to achieve separation.

B. The gas centrifuge process uses a large number of rotating cylinders in series and parallel configurations. Gas is introduced and rotated at high speed, concentrating the component of higher molecular weight toward the outer wall of the cylinder and the component of lower molecular weight toward the center. The enriched and the depleted gases are removed by scoops.

Nuclear fuel is made to maintain both its chemical and physical properties under the extreme conditions of heat and radiation present inside an operating reactor vessel. After the UF_6 is chemically converted to UO_2, the powder is blended, milled, pressed, and fused into ceramic fuel pellets about the size of a fingertip. The pellets are stacked into tubes about 14 feet (2.6 meters) long made of material called "cladding" (such as zirconium alloys). After careful inspection, the resulting fuel rods are bundled into fuel assemblies for use in reactors. The assemblies are washed, inspected, and stored in a special rack until ready for shipment to a nuclear power plant site. The NRC inspects this operation to ensure it is conducted safely.

Domestic Safeguards Program

The NRC's domestic safeguards program for fuel cycle facilities and transportation is aimed at ensuring that special nuclear material (such as plutonium or enriched uranium) is not stolen for possible malevolent uses. The program also works to ensure that such material does not pose an unreasonable risk to the public from sabotage or terrorism. The NRC verifies through licensing and inspection activities that licensees apply safeguards to protect special nuclear material. Additionally, the NRC and DOE developed the Nuclear Materials Management and Safeguards System (NMMSS) to track transfers and inventories of special nuclear material, source material from abroad, and other material. The NRC has issued licenses to approximately 180 facilities authorizing them to possess special nuclear material in quantities ranging from a single kilogram to multiple tons. These licensees verify and document their inventories in the NMMSS database. The NRC or State governments license several hundred additional sites that possess special nuclear material in smaller quantities (typically ranging from 1 gram to tens of grams). Licensees that possess small amounts of special nuclear material are now required to confirm their inventory annually in the NMMSS database. Previously, those licensees reported transfers of material but not annual inventories.

Figure 39. Simplified Fuel Fabrication Process

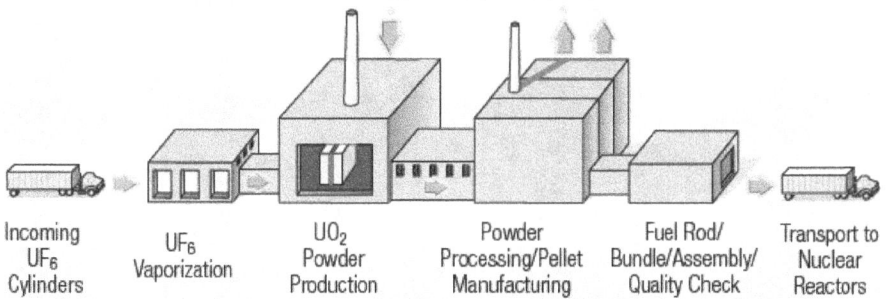

| Incoming UF$_6$ Cylinders | UF$_6$ Vaporization | UO$_2$ Powder Production | Powder Processing/Pellet Manufacturing | Fuel Rod/ Bundle/Assembly/ Quality Check | Transport to Nuclear Reactors |

Fabrication of commercial light-water reactor fuel consists of the following three basic steps:

(1) the chemical conversion of UF$_6$ to UO$_2$ powder

(2) a ceramic process that converts UO$_2$ powder to small ceramic pellets

(3) a mechanical process that loads the fuel pellets into rods and constructs finished fuel assemblies

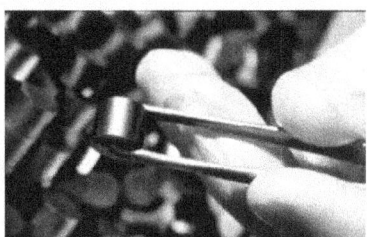

Small ceramic fuel pellets.

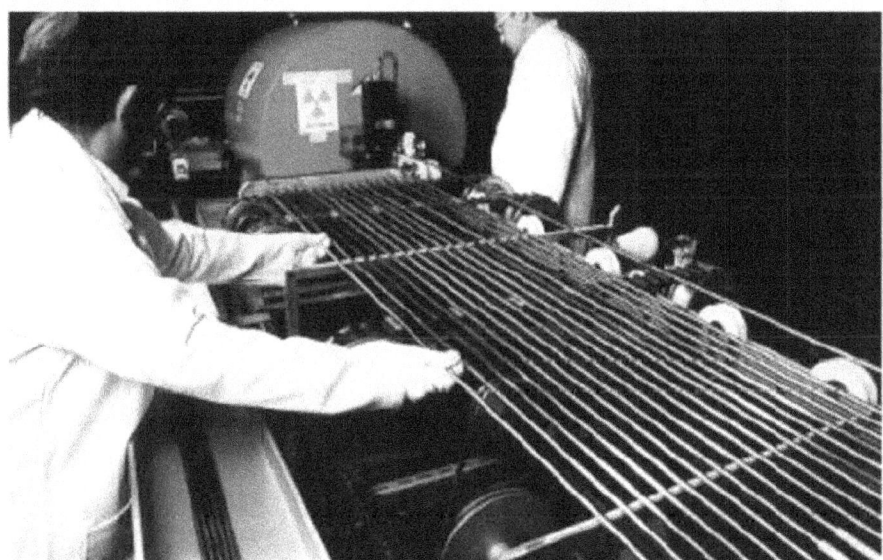

Fuel pellets being assembled into fuel rods.

Radioactive Waste

Spent Fuel Pool

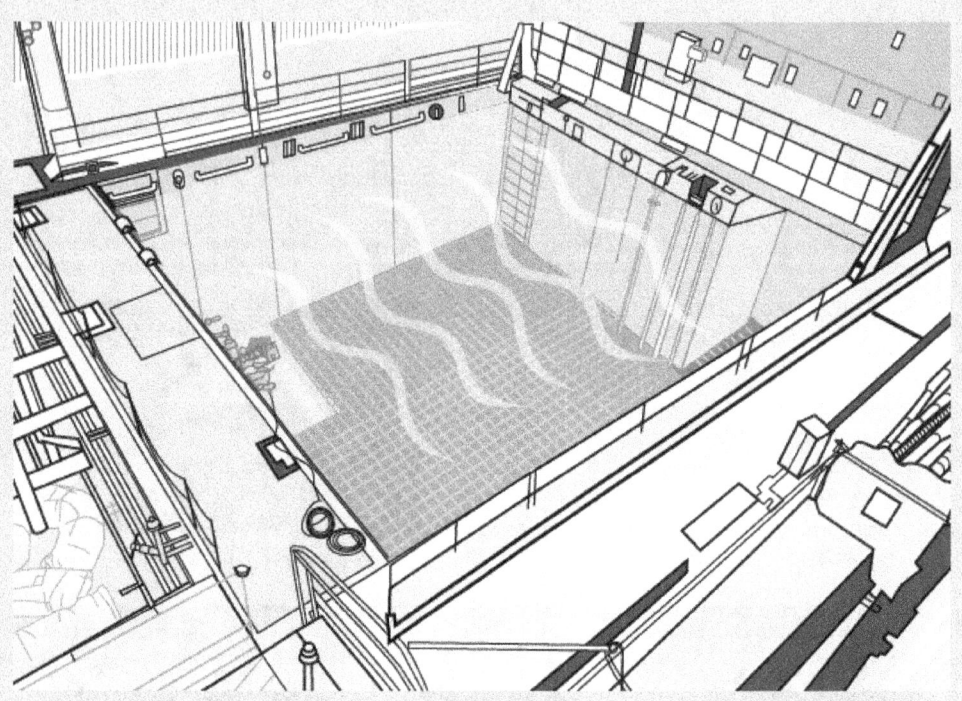

Facilities Undergoing Decommissioning

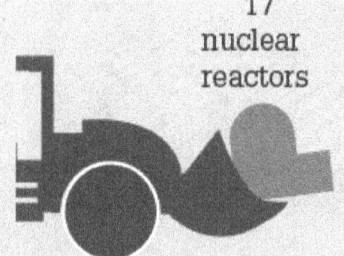

17
nuclear
reactors

18
complex
material
sites

11
research
and test
reactors

1
fuel
cycle
facility

11
uranium
recovery
facilities

Spent Fuel Storage Cask

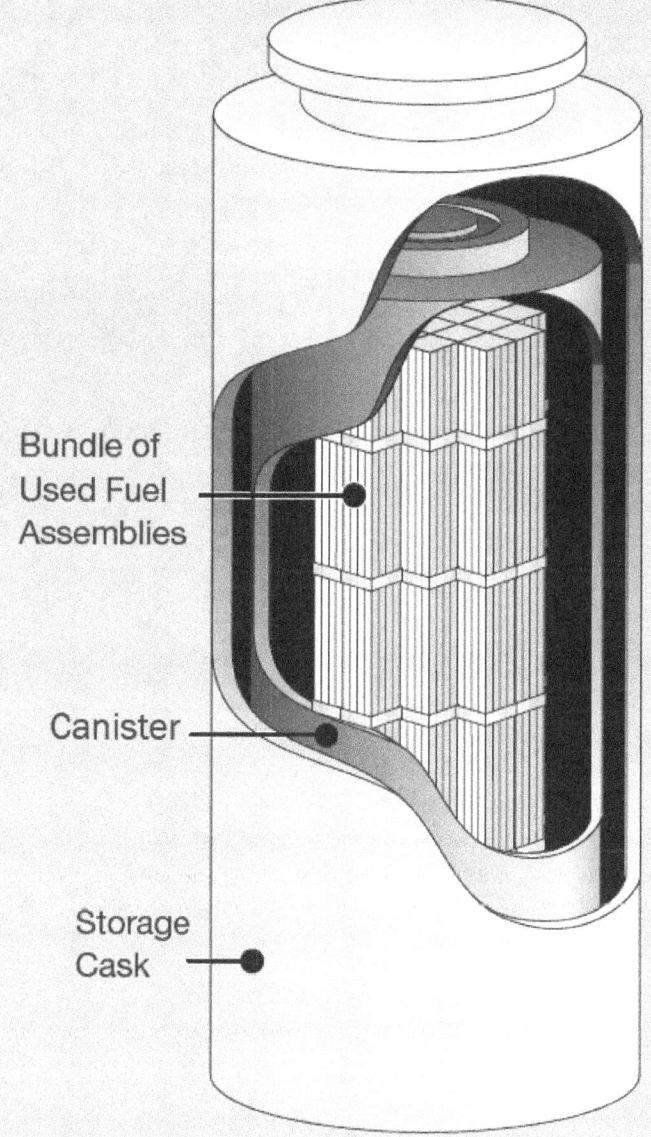

Bundle of Used Fuel Assemblies

Canister

Storage Cask

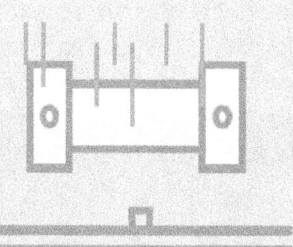

ENSURE SAFE SHIPPING CONTAINERS

Low-Level Radioactive Waste Disposal

Low-level radioactive waste (LLW) includes items that have become contaminated with radioactive material or have become radioactive through exposure to neutron radiation. This waste typically consists of contaminated protective shoe covers and clothing, wiping rags, mops, filters, reactor water treatment residues, equipment and tools, medical tubes, swabs, injection needles, syringes, and laboratory animal carcasses and tissue.

The radioactivity can range from just-above-background levels found in nature to very high levels from the parts inside the reactor vessel in a nuclear power plant. Licensees store some lower level radioactive waste on site until it has decayed and lost its radioactivity. Then it can be disposed of as ordinary trash. Waste that does not decay fairly quickly is stored until amounts are large enough for shipment to an LLW disposal site in containers approved by DOT and the NRC.

Commercial LLW is disposed of in facilities licensed by either the NRC or Agreement States in accordance with health and safety requirements. The facilities are designed, constructed, and operated to meet safety standards. The operator of the facility also extensively characterizes the site on which the facility is located and analyzes how the facility will perform in the future. Current LLW disposal uses shallow land disposal sites with or without concrete vaults (see Figure 40).

The NRC has developed a classification system for LLW based on its potential hazards. It has specified disposal and waste requirements for each of the three classes of waste—Classes A, B, and C—that are acceptable for disposal in near-surface facilities. These classes have progressively higher levels of concentrations of radioactive material, with A having the lowest and C having the highest level. Class A waste accounts for approximately 96 percent of the total volume of LLW. Determination of the classification of waste is a complex process. A fourth class of LLW, greater than Class C, is not generally acceptable for near-surface, shallow-depth disposal. By law, DOE is responsible for disposal of greater than Class C waste under an NRC license.

The volume and radioactivity of waste vary from year to year based on the types and quantities of waste shipped each year. Waste volumes currently include several million cubic feet each year from reactor facilities undergoing decommissioning and from cleanup of contaminated sites.

Figure 40. Low-Level Waste Disposal

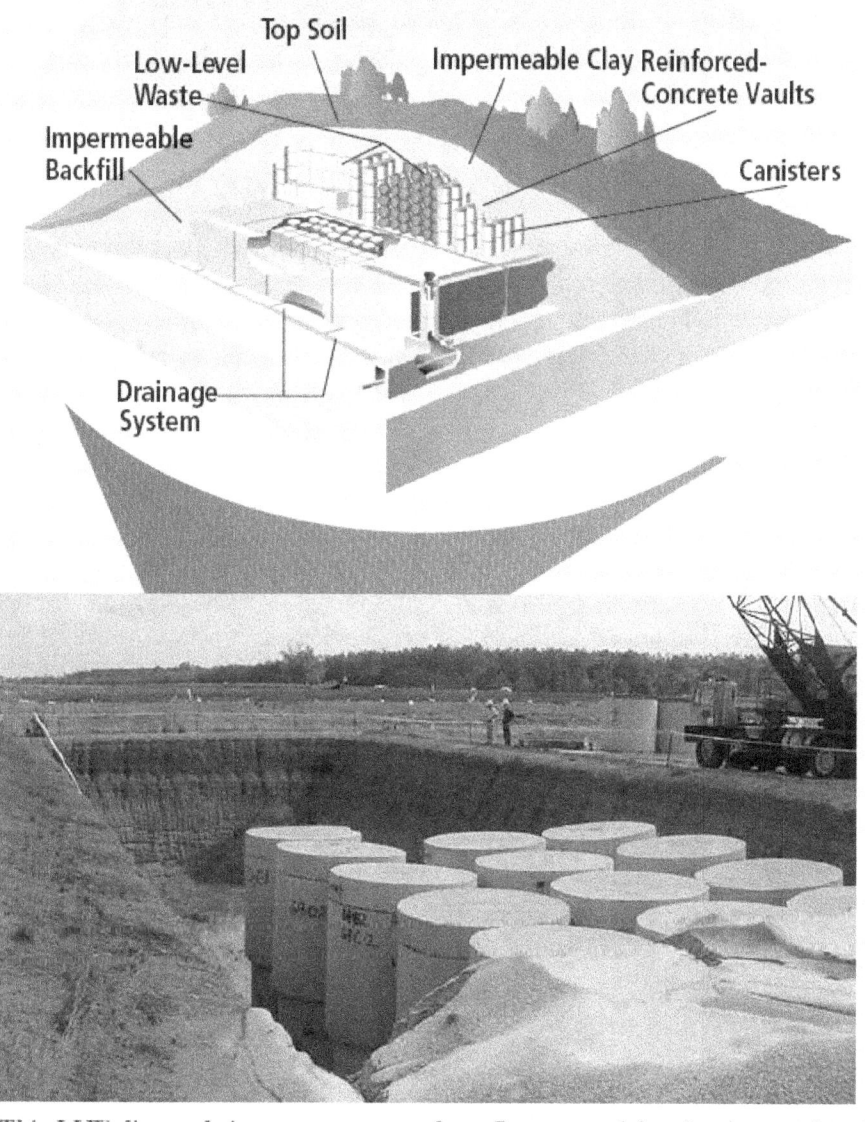

This LLW disposal site accepts waste from States participating in a regional disposal agreement.

The Low-Level Radioactive Waste Policy Amendments Act of 1985 gave the States responsibility for the disposal of LLW. The Act authorized States to do the following:

- Form regional compacts, with each compact to provide for LLW disposal site access.

- Manage LLW import to, and export from, a compact.

See Appendix P for Regional Compacts

- Exclude waste generated outside a compact.

The States have licensed four active LLW disposal facilities:

- Barnwell, located in Barnwell, SC—Previously, Barnwell accepted waste from all U.S. generators. As of July 2008, Barnwell accepts waste from the Atlantic Compact States (Connecticut, New Jersey, and South Carolina). The State of South Carolina licenses Barnwell to receive Classes A, B, and C of LLW.

- EnergySolutions, located in Clive, UT—EnergySolutions accepts waste from all regions of the United States. Utah licenses EnergySolutions for Class A waste only.

- Hanford, located in Hanford, WA—Hanford accepts waste from the Northwest Compact States (Alaska, Hawaii, Idaho, Montana, Oregon, Utah, Washington, Wyoming) The Rocky Mountain Compact States (Colorado, Nevada and New Mexico). The State of Washington licenses Hanford to receive Classes A, B, and C of LLW.

- Waste Control Specialist (WCS), located in Andrews, TX—The State of Texas licensed WCS in 2009 to receive Classes A, B, and C of LLW from the Texas Compact, which consists of Texas and Vermont. WCS is receiving LLW as of 2011.

Closed LLW disposal facilities licensed by the NRC or Agreement States include the following:

- Beatty, NV, closed 1993

- Sheffield, IL, closed 1978

- Maxey Flats, KY, closed 1977

- West Valley, NY, closed 1975

High-Level Radioactive Waste Management

Spent Nuclear Fuel Storage

Commercial spent nuclear fuel, although highly radioactive, is stored safely and securely in 35 States. This includes 31 States with operating nuclear power reactors, where spent fuel is safely stored on site in spent fuel pools and in some dry casks. The remaining four States—Colorado, Idaho, Maine, and Oregon—do not have operating power reactors but are safely storing spent fuel at storage facilities. Waste can be stored safely in pools or casks for 100 years or more.

As of January 2012, the amount of commercial spent fuel in safe storage at commercial nuclear power plants was an estimated 67,000 metric tons. The amount of spent fuel in storage at commercial nuclear power plants is expected to increase at a rate of approximately 2,000 metric tons per year. The NRC licenses and regulates the storage of spent fuel, both at commercial nuclear power plants and at storage facilities located away from reactors.

See Appendix N and O for dry spent fuel storage and licensees information.

Most reactor facilities were not designed to store the full amount of spent fuel that the reactor would generate during its operational life. Facilities originally planned to store spent fuel temporarily in deep pools of continuously circulating water that cools the spent fuel assemblies and provides shielding from radiation. After a few years, the facilities were expected to send the spent fuel to a recycling plant. However, in 1977, the Federal Government declared a moratorium on recycling spent fuel in the United States. Although the ban was later lifted, recycling has not been pursued.

To cope with the spent fuel they were generating, facilities expanded their storage capacity by using high-density storage racks in their spent fuel pools (see Figure 42). However, spent fuel pools are not a permanent storage solution. To provide supplemental storage, a portion of spent fuel inventories is stored in dry casks on site. These facilities are called independent spent fuel storage installations (ISFSIs) and are licensed by the NRC. These large casks are typically made of leak-tight, welded, and bolted steel and concrete surrounded by another layer of steel or concrete. The spent fuel sits in the center of the nested canisters in an inert gas. Dry cask storage shields people and the environment from radiation and keeps the spent fuel inside dry and nonreactive (see Figure 43).

Figure 41. Spent Fuel Generation and Storage after Use

1 A nuclear reactor is powered by enriched uranium-235 fuel. Fission (splitting of atoms) generates heat, which produces steam that turns turbines to produce electricity. A reactor rated at several hundred megawatts may contain 100 or more tons of fuel in the form of bullet-sized pellets loaded into long metal rods that are bundled together into fuel assemblies. PWRs contain between 150 and 200 fuel assemblies. BWRs contain between 370 and 800 fuel assemblies.

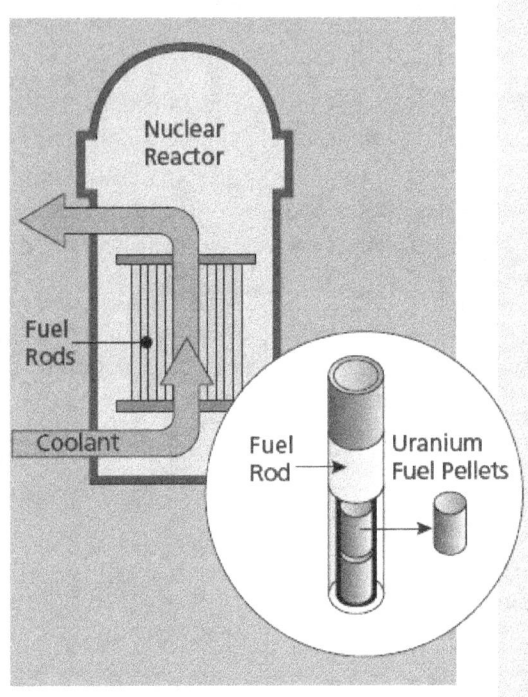

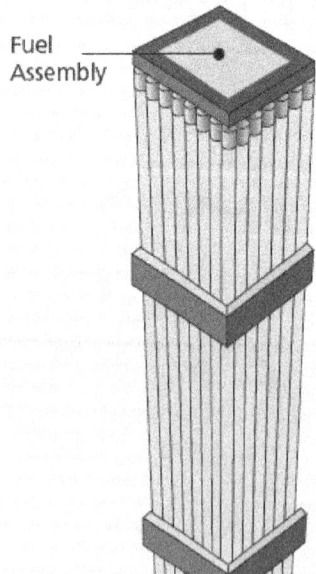

2 After about 6 years, spent fuel assemblies—typically 14 feet (4.3 meters) long and containing nearly 200 fuel rods for PWRs and 80–100 fuel rods for BWRs—are removed from the reactor and allowed to cool in storage pools for a few years. At this point, the 900-pound (409-kilogram) assemblies contain only about one-fifth the original amount of uranium-235.

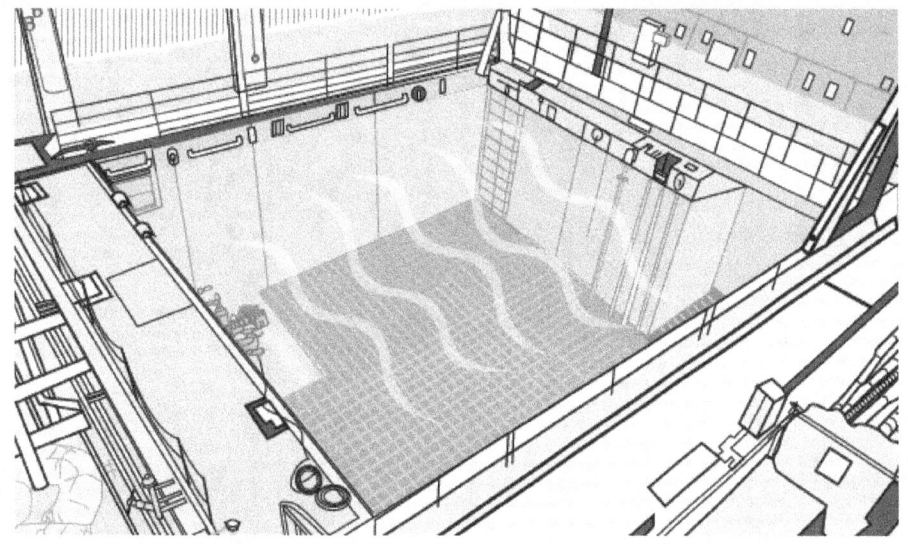

3 Commercial light-water nuclear reactors store spent radioactive fuel in a steel-lined, seismically designed concrete pool under about 40 feet (12.2 meters) of water that provides shielding from radiation. Water pumps supply continuously flowing water to cool the spent fuel. Extra water for the pool is provided by other pumps that can be powered from an onsite emergency diesel generator. Support features, such as water-level monitors and radiation detectors, are also in the pool. Spent fuel is stored in the pool until it can be transferred to dry casks on site (as shown in Figure 42) or transported off site to a high-level radioactive waste disposal site.

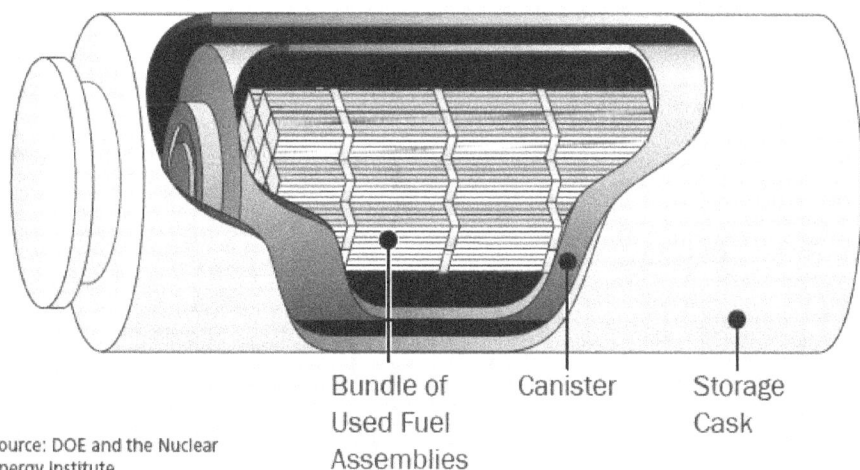

Bundle of
Used Fuel
Assemblies

Canister

Storage
Cask

Source: DOE and the Nuclear
Energy Institute

Figure 42. Dry Storage of Spent Nuclear Fuel

At some nuclear reactors across the country, spent fuel is kept on site, typically above ground, in systems basically similar to the ones shown here.

1 *Once the spent fuel has sufficiently cooled, it is loaded into special canisters that are designed to hold nuclear fuel assemblies. Water and air are removed. The canister is filled with inert gas, welded shut, and rigorously tested for leaks. It is then placed in a cask for storage or transportation. The NRC has approved the storage of up to 40 PWR assemblies and up to 68 BWR assemblies in each canister. The dry casks are then loaded onto concrete pads.*

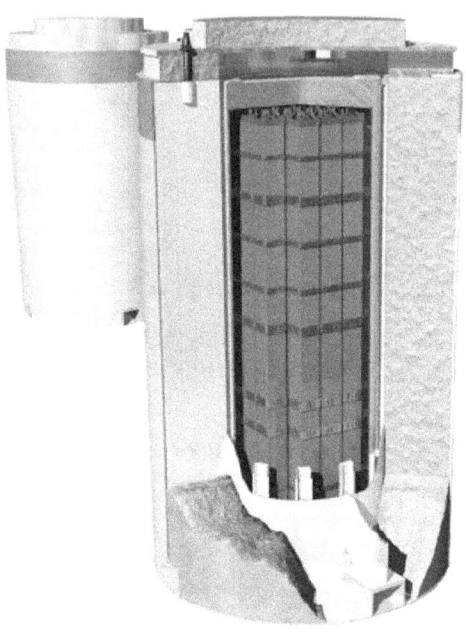

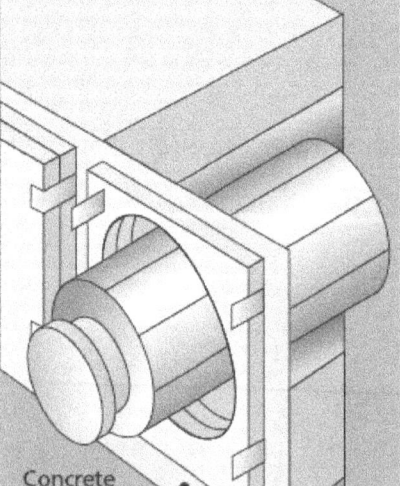

Concrete
Storage
Bunker

2 *The canisters can also be stored in aboveground concrete bunkers, each of which is about the size of a one-car garage.*

Currently, there are 65 licensed ISFSIs in the United States. As of 2012, NRC-licensed ISFSIs were storing spent fuel in over 1,161 loaded dry casks (see Figure 44). The NRC authorizes storage of spent fuel at an ISFSI under two licensing options: (1) site-specific licensing and (2) general licensing. Site-specific licenses granted by the NRC after a safety review contain technical requirements and operating conditions for the ISFSI and specify what the licensee is authorized to store at the site. The initial and renewal license terms for an ISFSI are not to exceed 40 years from the date of issuance. A general license from the NRC authorizes a licensee who operates a nuclear power reactor to store spent fuel on site in dry storage casks. Under the general license, the authority to use a storage cask is tied to the cask's certificate of compliance (CoC) term. A CoC is issued to the cask vendor through rulemaking. Several dry storage cask designs have received certificates. Initial and renewed CoCs are issued for terms not to exceed 40 years.

At least 30 days before the certificate expiration date, the cask vendor may apply for renewal. If the cask vendor does not apply for renewal, a general licensee may apply for renewal. Before using the cask, general licensees must certify that the cask meets the conditions in the certificate, that the concrete pads under the casks can adequately support the loads, and that the levels of radiation from the casks meet NRC standards. Specific license and CoC renewal applications must include an analysis that considers the effects of aging on structures, systems, and components of safety for the requested renewal term.

Public Involvement

The public can participate in decisions about spent fuel storage, as it can in many licensing and rulemaking decisions. The Atomic Energy Act of 1954, as amended, and the NRC's own regulations provide the opportunity for public hearings for site-specific licensing actions and allow for public comments on certificate rulemakings. Interested members of the public may also file petitions for rulemaking. Additional information on ISFSIs is available on the NRC Web site (see the Web Link Index).

Spent Nuclear Fuel Disposal

The current U.S. policy governing permanent disposal of high-level radioactive waste is defined by the Nuclear Waste Policy Act of 1982, as amended, and the Energy Policy Act of 1992. These acts specify that high-level radioactive waste will be disposed of underground in a deep geologic repository.

Figure 43. Licensed/Operating Independent Spent Fuel Storage Installations by State

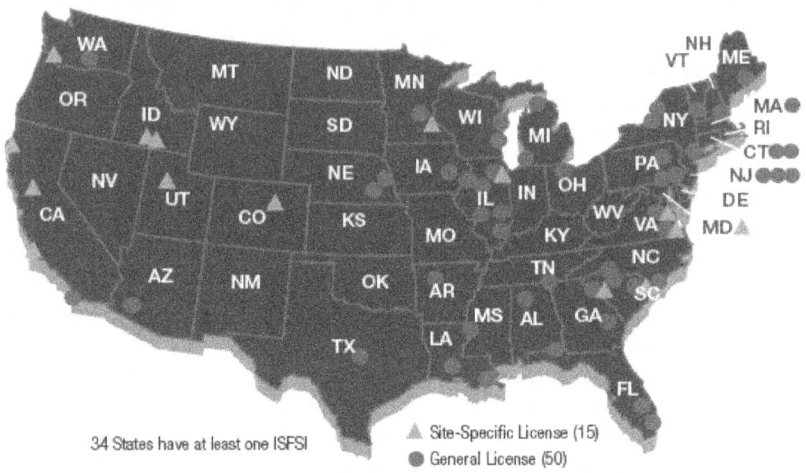

34 States have at least one ISFSI

▲ Site-Specific License (15)
● General License (50)

ALABAMA
- ● Browns Ferry
- ● Farley

ARIZONA
- ● Palo Verde

ARKANSAS
- ● Arkansas Nuclear

CALIFORNIA
- ▲ Diablo Canyon
- ▲ Rancho Seco
- ● San Onofre
- ▲ Humboldt Bay

COLORADO
- ▲ Fort St. Vrain

CONNECTICUT
- ● Haddam Neck
- ● Millstone

FLORIDA
- ● St. Lucie
- ● Turkey Point

GEORGIA
- ● Hatch

IDAHO
- ▲ DOE: TMI-2 (Fuel Debris)
- ▲ Idaho Spent Fuel Facility

ILLINOIS
- ● Braidwood
- ● Byron
- ▲ GE Morris (Wet)
- ● Dresden
- ● La Salle
- ● Quad Cities

IOWA
- ● Duane Arnold

LOUISIANA
- ● River Bend
- ● Waterford

MAINE
- ● Maine Yankee

MARYLAND
- ▲ Calvert Cliffs

MASSACHUSETTS
- ● Yankee Rowe

MICHIGAN
- ● Big Rock Point
- ● Palisades

MINNESOTA
- ● Monticello
- ▲ Prairie Island

MISSISSIPPI
- ● Grand Gulf

NEBRASKA
- ● Cooper
- ● Ft. Calhoun

NEW HAMPSHIRE
- ● Seabrook

NEW JERSEY
- ● Hope Creek
- ● Salem
- ● Oyster Creek

NEW YORK
- ● Indian Point
- ● FitzPatrick
- ● Ginna

NORTH CAROLINA
- ● Brunswick
- ● McGuire

OHIO
- ● Davis-Besse

OREGON
- ▲ Trojan

PENNSYLVANIA
- ● Limerick
- ● Susquehanna
- ● Peach Bottom

SOUTH CAROLINA
- ●▲ Oconee
- ●▲ Robinson
- ● Catawba

TENNESSEE
- ● Sequoyah

TEXAS
- ● Comanche Peak

UTAH
- ▲ Private Fuel Storage

VERMONT
- ● Vermont Yankee

VIRGINIA
- ●▲ Surry
- ●▲ North Anna

WASHINGTON
- ● Columbia

WISCONSIN
- ● Point Beach
- ● Kewaunee**

DOE submitted its license application to the NRC on June 3, 2008, for Yucca Mountain in Nevada. The NRC formally accepted it for review in September 2008 and began the detailed technical review and associated adjudicatory activities. In 2009, President Barack Obama announced that the administration would terminate the Yucca Mountain program while developing a disposal alternative.

In September 2011, the NRC completed an orderly closure of its Yucca Mountain activities. As part of the orderly closure, the NRC prepared three technical evaluation reports, in addition to one volume of a safety evaluation report published earlier. The NRC also prepared 46 additional reports that capture important technical or regulatory information, insights, and lessons learned from more than 25 years of work during the prelicensing and licensing phases of the Yucca Mountain project.

On January 29, 2010, President Obama directed the Secretary of Energy to establish the Blue Ribbon Commission on America's Nuclear Future (BRC) to conduct a comprehensive review of policies for managing the back end of the nuclear fuel cycle and recommend a new strategy. The BRC provided its final recommendations to the Secretary of Energy on January 26, 2012. Several of the BRC's recommendations are related to ongoing areas of NRC regulatory activities. The key areas in this effort are the nuclear fuel cycle, spent fuel storage and transportation, and high-level waste disposal.

Reprocessing

In the United States, spent nuclear fuel is stored safely and securely either at a nuclear power plant or at a storage facility away from a plant. Some countries reprocess their spent nuclear fuel to recover fissile material and use it to generate more energy. Although the NRC has not received an application for a reprocessing facility, the agency has completed an initial analysis of the existing regulatory framework in preparation for such an application. The NRC has developed a draft technical basis document for revising the existing regulatory framework for reprocessing facilities, to ensure that a potential commercial reprocessing facility can be licensed efficiently and effectively and operate safely.

Transportation

The NRC is also involved in the transportation of spent nuclear fuel. It establishes safety criteria for spent fuel shipping casks and certifies cask designs. Casks are designed to meet the following safety criteria under both normal and accident conditions:

- prevent the loss or dispersion of radioactive contents;

- provide shielding and heat dissipation; and

- prevent nuclear criticality (a self-sustaining nuclear chain reaction).

Spent fuel shipping casks must be designed to survive a sequence of tests, including a 9-meter (30-foot) drop onto an unyielding surface, a puncture test, and a fully engulfing fire at 1,475 degrees Fahrenheit (802 degrees Celsius) for 30 minutes. This very severe test sequence, akin to the cask striking a concrete pillar along a highway at a high speed and being engulfed in a very severe and long-lasting fire, simulates conditions more severe than 99 percent of vehicle accidents (see Figure 45).

Principal Licensing and Inspection Activities

The NRC regulates spent fuel transportation through a combination of safety and security requirements, certification of transportation casks, inspections, and a system of monitoring to ensure that requirements are being met. Specifically, each year, the NRC does the following:

- conducts about 1,000 transportation safety inspections of fuel, reactor, and materials licensees;

- reviews, evaluates, and certifies approximately 80 new, renewal, or amended transport package design applications;

- inspects about 28 dry storage and transport package licensees; and

- reviews and evaluates approximately 150 license applications for the import or export of nuclear materials.

Additional information on materials transportation is available on the NRC Web site (see the Web Link Index).

Figure 44. Ensuring Safe Spent Fuel Shipping Containers

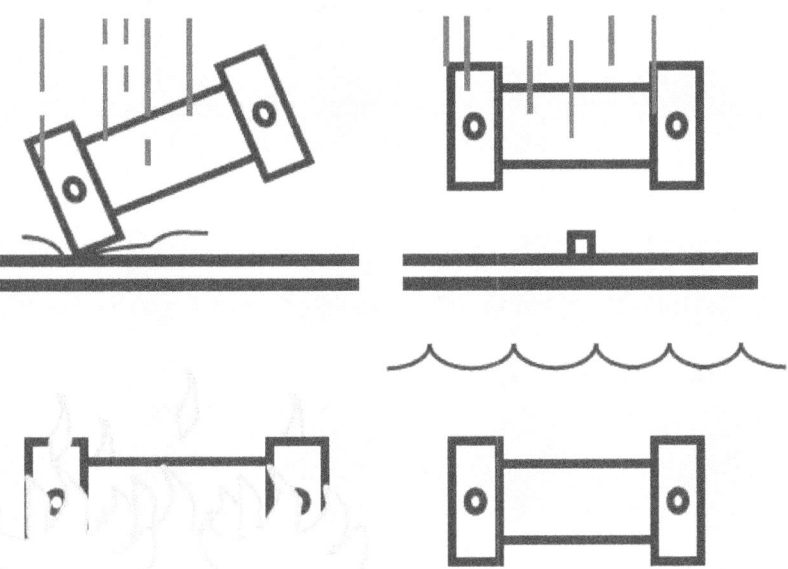

The impact (free drop and puncture), fire, and water-immersion tests are considered in sequence to determine their cumulative effects on a given package.

Decommissioning

Decommissioning is the safe removal of a nuclear facility from service and the reduction of residual radioactivity to a level that permits release of the property and termination of the license. The NRC rules establish site-release criteria and provide for unrestricted and, under certain conditions, restricted release of a site. The NRC rules also require licensees authorized to possess radioactive materials above a threshold amount to maintain financial assurance that funds will be available when needed for decommissioning.

The NRC regulates the decontamination and decommissioning of materials and fuel cycle facilities, nuclear power plants, research and test reactors, and uranium recovery facilities, with the ultimate goal of license termination. The NRC terminates approximately 150 materials licenses each year. Most of these license terminations are routine, and the sites require little, if any, remediation to meet the NRC's release criteria for unrestricted access. The decommissioning program focuses on the termination of licenses that are not routine, because the sites involve more complex decommissioning activities (see Figure 45).

As of early April 2012, the following facilities were undergoing decommissioning (see Figure 46) under NRC jurisdiction:

- 17 nuclear power and early demonstration reactors

- 18 complex material sites

- 11 research and test reactors

- 1 fuel cycle facility

- 11 uranium recovery facilities

See Appendices B, J and R for licensees undergoing decommissioning.

The "Status of the Decommissioning Program 2011 Annual Report" provides additional information on the decommissioning programs of the NRC and Agreement States. More information is on the NRC Web site (see the Web Link Index).

As part of the decommissioning process, the cooling tower of a nuclear power plant is imploded.

Figure 45. Locations of NRC-Regulated Complex Material Sites Undergoing Decommissioning

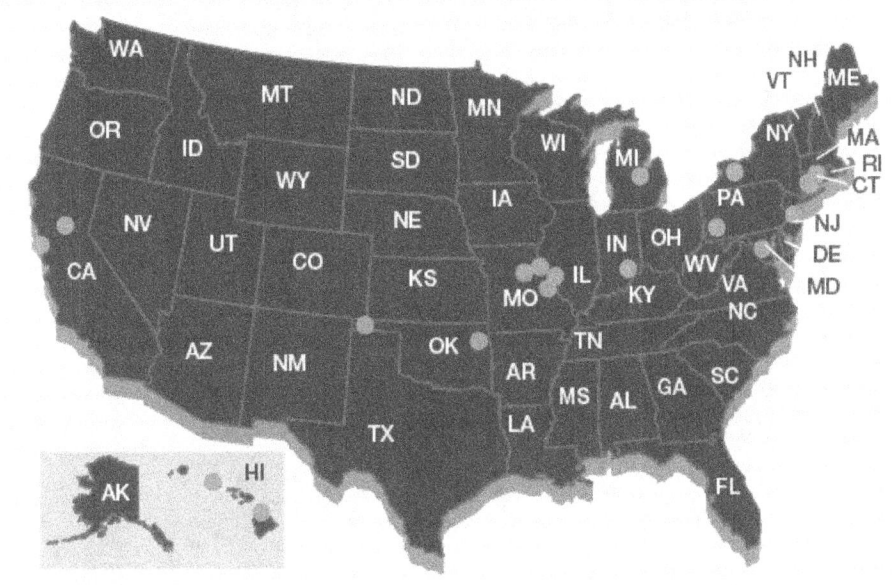

● NRC-regulated complex material sites (18)

Figure 46. Facilities Undergoing Decommissioning Under NRC Jurisdiction

| 17 nuclear reactors | 18 complex material sites | 11 research and test reactors | 1 fuel cycle facility | 11 uranium recovery facilities |

Security and Emergency Preparedness

SECURITY COMPONENTS

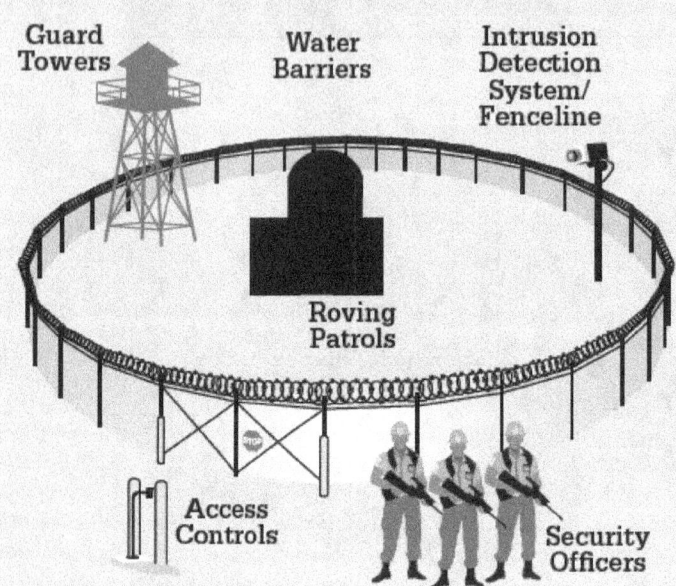

Guard Towers

Water Barriers

Intrusion Detection System/ Fenceline

Roving Patrols

Access Controls

Security Officers

Emergency Planning Zones

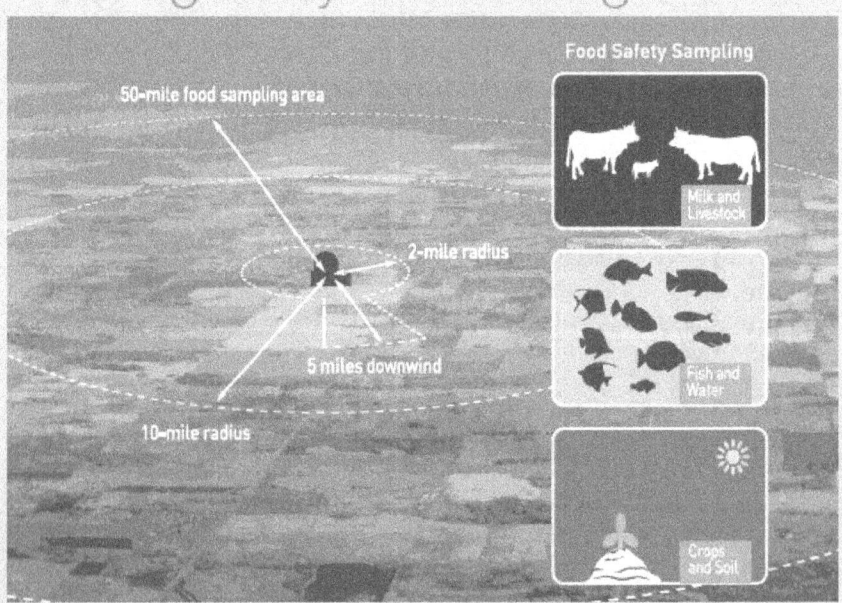

Food Safety Sampling

50-mile food sampling area

2-mile radius

5 miles downwind

10-mile radius

Milk and Livestock

Fish and Water

Crops and Soil

94

Emergency Classifications

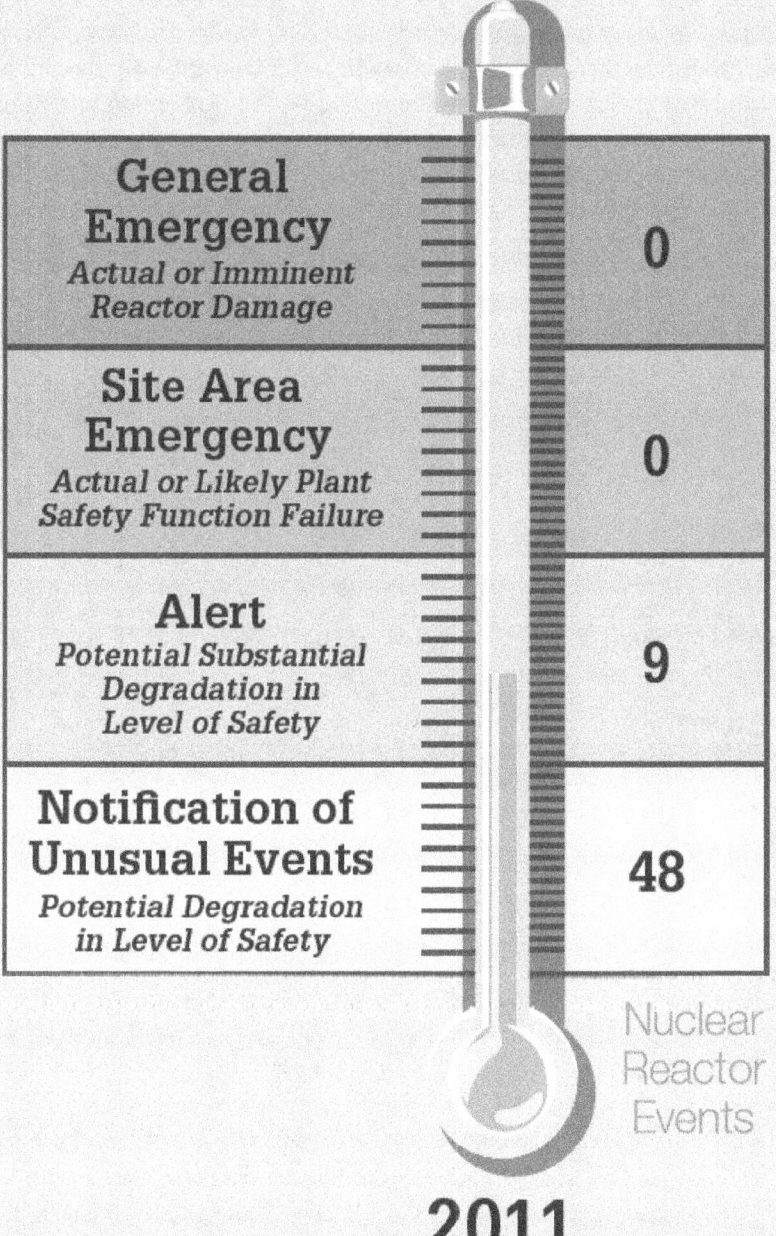

General Emergency *Actual or Imminent Reactor Damage*	0
Site Area Emergency *Actual or Likely Plant Safety Function Failure*	0
Alert *Potential Substantial Degradation in Level of Safety*	9
Notification of Unusual Events *Potential Degradation in Level of Safety*	48

Nuclear Reactor Events

2011

Overview

Nuclear security is a high priority for the NRC. For the past several decades, effective NRC regulation and strong partnerships with a variety of Federal, State, Tribal, and local authorities have ensured effective implementation of security programs at the Nation's nuclear facilities and radioactive materials across the country. In fact, nuclear power plants are likely the best protected private sector facilities in the United States. However, given today's threat environment, the agency recognizes the need for continued vigilance and high levels of security.

In recent years, the NRC has made many enhancements to bolster the security of nuclear power plants. Because nuclear power plants are inherently robust structures, these additional security upgrades largely focus on the following improvements (see Figure 47):

* well-trained and armed security officers;
* high-tech equipment and physical barriers;
* greater standoff distances for vehicle checks;
* intrusion detection and surveillance systems;
* tested emergency preparedness and response plans; and
* restrictive site access control, including background checks and fingerprinting of workers.

Additional layers of security are provided by coordinating and sharing threat information among DHS, the U.S. Federal Bureau of Investigation, intelligence agencies, the U.S. Department of Defense, and local law enforcement.

Facility Security

In accordance with NRC regulations, nuclear power plants and Category I fuel facilities must be able to defend successfully against a set of hypothetical threats that the agency calls the design-basis threat (DBT). This includes threats that challenge a plant's physical security, personnel security, and cyber security. The NRC does not make details of the DBT public because of security concerns. However, the agency continuously evaluates this set of hypothetical threats against real-world intelligence to ensure that the DBT remains current. To test the adequacy of a nuclear power plant's defenses against the DBT, the NRC conducts rigorous "force-on-force" inspections.

Figure 47. Security Components

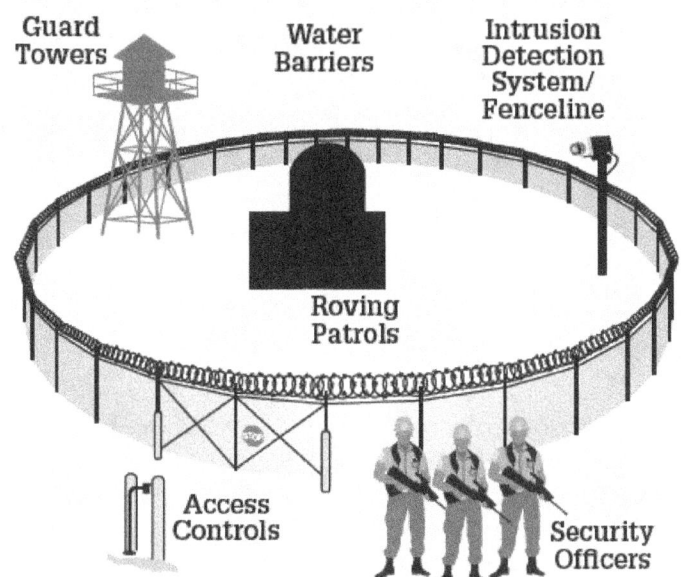

Guard Towers

Water Barriers

Intrusion Detection System/ Fenceline

Roving Patrols

Access Controls

Security Officers

Protecting nuclear facilities requires all the security features to come together and work as one.

No Trespassing
Armed Security Officers on Duty

Deadly Force
Could be Authorized

Code of Federal Regulations 10 CFR 73.55 (b)(3)

Licensees are authorized to use deadly force while protecting nuclear facilities from intruders.

Access control security gates within a nuclear facility provide another layer of protection.

During these inspections, exercises are conducted in which a highly trained mock adversary force "attacks" a nuclear facility. Beginning in 2004, the NRC began conducting more challenging and realistic force-on-force exercises that also occur more frequently. To ensure that facilities meet their security requirements, the NRC inspects nuclear power plants and fuel fabrication facilities, spending about 16,000 hours a year scrutinizing security including 8,000 hours of force-on-force inspections). Publicly available portions of security-related inspection reports can be found on the NRC Web site (see the Web Link Index). Inspection reports are not available for the NRC-licensed highly enriched uranium fuel facilities.

Cyber Security

Nuclear facilities use digital and analog systems to monitor, control, and run various types of equipment and to obtain and store vital information. Protecting these systems and the information they contain from sabotage or malicious use is called "cyber security." However, nuclear plant reactor control systems are isolated from the internet to protect reactors. All nuclear power plants licensed by the NRC must have a cyber security program. A new cyber security rule, issued in 2009, requires each nuclear power facility to submit a cyber security plan and implementation timeline for NRC approval. Once the licensee has fully implemented its program, the NRC will conduct a comprehensive inspection on site. The NRC has formed a cyber security team that includes technology and threat experts who constantly evaluate and identify emerging cyber-related issues that could affect plant systems. This team makes recommendations to other NRC offices and programs on cyber security issues.

Materials Security

The security of radioactive materials is important for a number of reasons. For example, terrorists could use radioactive materials to make a radiological dispersal device such as a dirty bomb. The NRC works with its Agreement States, other Federal agencies, IAEA, and licensees to protect radioactive material from theft or diversion. The agency has made improvements and upgrades to the joint NRC-DOE database that tracks the movement and location of certain forms and quantities of special nuclear material. In early 2009, the NRC deployed its new NSTS, designed to track the most risk-sensitive sources on a continuous basis. Other improvements allow U.S. Customs and Border Protection agents to promptly validate whether radioactive materials coming into the United States are properly licensed by the NRC.

Emergency Preparedness

As a condition of their license, operators of nuclear facilities develop and maintain effective emergency plans and procedures. The NRC inspects licensees to ensure that they are prepared to deal with emergencies. In addition, the agency monitors performance indicators related to emergency preparedness (see Figure 48).

Well-developed and practical emergency preparedness plans ensure that a nuclear power plant operator can protect public health and safety in the unlikely event of an emergency.

See Appendix H for lists of industry performance indicators.

The NRC staff participates in emergency preparedness exercises, some of which include security- and terrorism-based scenarios. To form a coordinated system of emergency preparedness and response, as part of these exercises, the NRC works with licensees; Federal agencies; State, Tribal, and local officials; and first responders. This system includes public information, preparations for evacuation, instructions for sheltering, and other actions to protect the residents near nuclear power plants in the event of a serious incident.

The NRC assesses the ability of nuclear power plant operators to respond to emergencies. For nuclear power plants, operators are required to conduct full-scale exercises with the NRC, the Federal Emergency Management Agency (FEMA), and State and local officials at least once every 2 years. These exercises

Figure 48. Industry Performance Indicators: FYs 2002–2011 Averages for 104 Plants

Alert and Notification System (ANS) Reliability

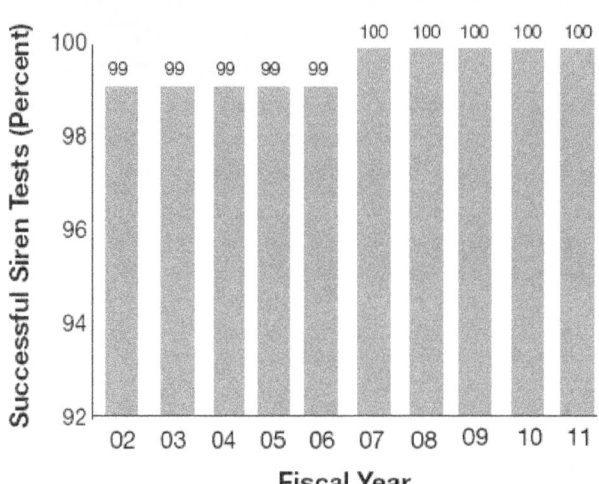

This shows the percentage of ANS sirens that successfully operated during periodic tests in the previous year. The result is an indicator of the reliability of the ANS to alert the public in an emergency.

Figure 49. Emergency Planning Zones

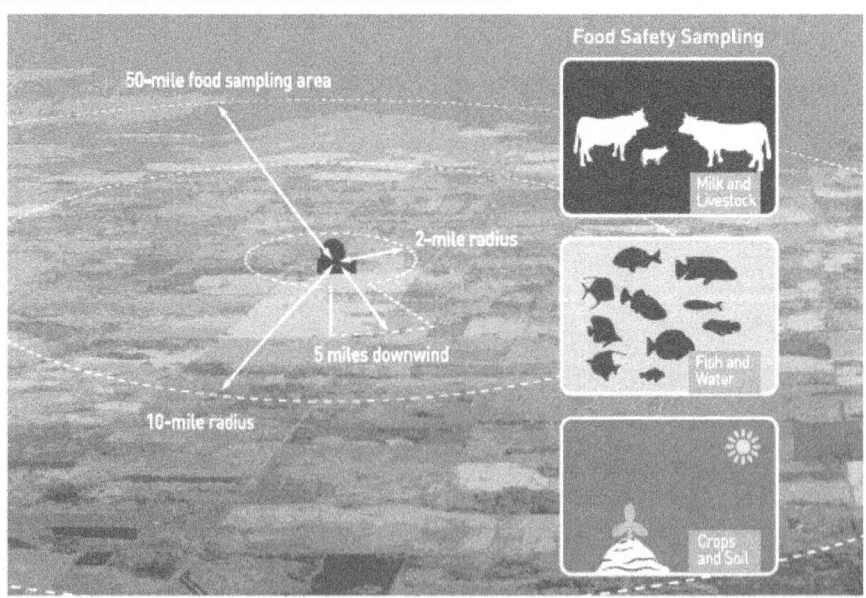

Note: A 2-mile ring around the plant is identified for evacuation along with a 5-mile zone downwind of the projected release path.

test and maintain the skills of the emergency responders and identify areas that need addressing. The NRC and FEMA evaluate these exercises. Between these 2-year exercises, nuclear power plant operators self-test their emergency plans in drills that NRC inspectors evaluate.

Emergency Planning Zones

Although emergency planning zones (EPZs) are meant to be expandable, as necessary, for planning purposes, the NRC defines two zones around each nuclear power plant. The exact size and configuration of the zones vary from plant to plant based on local emergency response needs and capabilities, population, land characteristics, access routes, and jurisdictional boundaries (see Figure 49 for a typical EPZ around a nuclear plant).

The two types of EPZs are as follows:

- The plume exposure pathway EPZ extends about 10 miles in radius around a plant. Its primary concern is the exposure of the public to, and the inhalation of, airborne radioactive contamination. Research has shown that the most significant impacts of an accident would be expected in the immediate vicinity of a plant, and any initial protective actions, such as evacuations or sheltering in place, should be focused there.

- The ingestion pathway EPZ extends about 50 miles in radius around a plant. Its primary concern is the ingestion of food and liquid that is contaminated by radioactivity.

See Glossary for radiation sources and exposure pathways.

Protective Actions

During an actual radiological event, the NRC would perform dose calculations using radiation dose projection models that analyze release paths from power reactors. The dose calculations would also take into account weather conditions to project radiation doses. The NRC would confer with appropriate State and county governments on its assessment results. Plant personnel would also provide assessments. State and local officials in communities within the EPZ have detailed plans to protect public health and safety in the event of a radiological release. These officials make the protective action decision, including evacuations, based on the assessments that they have received. See Figure 50 for dose regulatory limits.

Figure 50. Radiation Doses and Regulatory Limits

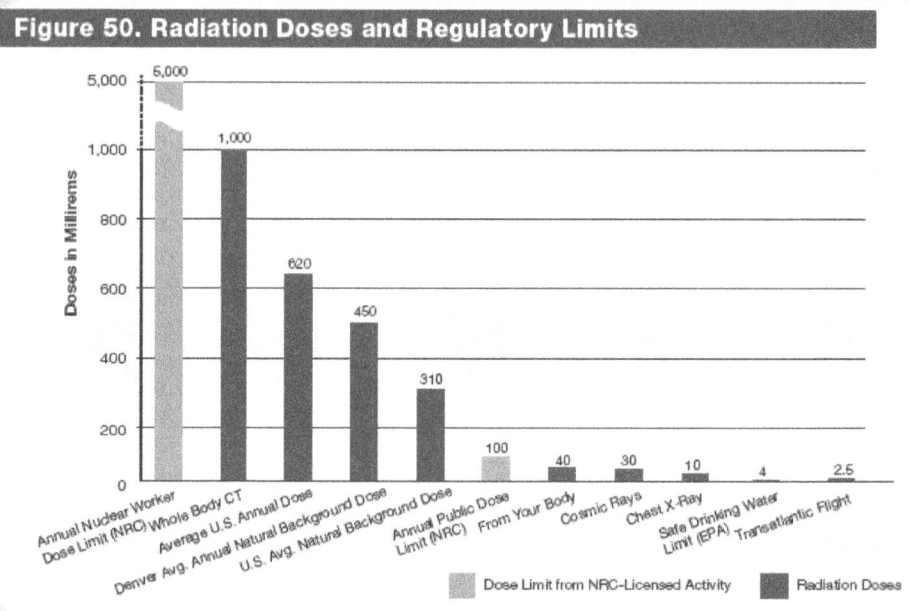

Doses in Millirems

Annual Nuclear Worker Dose Limit (NRC)	5,000
Whole Body CT	1,000
Average U.S. Annual Dose	620
Denver Avg. Annual Natural Background Dose	450
U.S. Avg. Natural Background Dose	310
Annual Public Dose Limit (NRC)	100
From Your Body	40
Cosmic Rays	30
Chest X-Ray	10
Safe Drinking Water Limit (EPA)	4
Transatlantic Flight	2.5

Dose Limit from NRC-Licensed Activity   Radiation Doses

NRC staff provided support to overseas counterparts during the Japan nuclear accident and examined available information to understand the implications for the United States.

Evacuation, Sheltering, and the Use of Potassium Iodine

Protective actions considered for a radiological emergency include evacuation, sheltering, and, as a supplement to these, the prophylactic use of potassium iodide (KI) to protect the thyroid from absorbing radioactive iodine. Under certain conditions, evacuation may be preferred to remove the public from further exposure to radioactive material. However, under some conditions, people may be instructed to take shelter in their homes, schools, or office buildings. Depending on the type of structure, sheltering can significantly reduce a person's dose compared to the dose received if he or she remained outside. In certain situations, KI is used as a supplement to sheltering.

Evacuation does not always call for the complete evacuation of the 10-mile zone around a nuclear power plant. In most cases, the release of radioactive material from a plant during a major incident would move with the wind, not in all directions surrounding the plant. The release would also expand and become less concentrated as it travels away from a plant. Therefore, evacuations can be planned to anticipate the path of the release.

Sheltering is a protective action that keeps people indoors to reduce exposure to radioactive material. It may be appropriate to shelter when the release of radioactive material is known to be short term or is controlled by the nuclear power plant operator. Additional information on emergency preparedness is available on the NRC Web site (see the Web Link Index).

Incident Response

Sharing information quickly among the NRC, other Federal and State agencies, and the nuclear industry is critical to responding promptly to any incident. The NRC staff supports several important Federal incident response centers that coordinate assessments of event-related information. The NRC Headquarters Operations Center, located in the agency's Headquarters in Rockville, MD, is staffed around the clock to disseminate information and coordinate response activities. To ensure the timely distribution of threat information, the NRC reviews intelligence reports and assesses suspicious activity.

The NRC works within the National Response Framework to respond to events. The framework guides the Nation in how to respond to complex events that may involve a variety of agencies and hazards.

The NRC expanded to 24-hour coverage in its operations center following the March 2011 earthquake and tsunami event in Japan.

Under this framework, the NRC retains its independent authority and ability to respond to emergencies that involve NRC-licensed facilities or materials. The NRC coordinates the Federal technical response to an incident that involves one of its licensees.

The NRC may request DHS support in responding to an emergency at an NRC-licensed facility or involving NRC-licensed materials. DHS may lead and manage the overall Federal response to an event, according to Homeland Security Presidential Directive 5, "Management of Domestic Incidents." In this case, the NRC would provide technical expertise and help share information among the various organizations and licensees.

In response to an incident involving possible releases of radioactive materials, the NRC activates its incident response program at its Headquarters Operations Center and one of its four Regional Incident Response Centers. Teams of specialists assemble at these Centers to evaluate event information and independently assess the potential impact on public health and safety. The NRC staff provides expert consultation, support, and assistance to State and local public safety officials and keeps the public informed of agency actions. Scientists and engineers at these Centers analyze the event and evaluate possible recovery strategies. Meanwhile, other NRC experts evaluate the effectiveness of protective actions the licensee has recommended that State and local officials implement. If needed, the NRC will dispatch a team of technical experts from the responsible regional office to the site of the incident. Augmenting the NRC's resident inspectors, who work at the plant, the team serves as the agency's onsite eyes and ears, allowing a firsthand assessment and face-to-face communications with all participants. The Headquarters Operations Center continues to provide around-the-clock Federal communications, logistical support, and technical analysis throughout the response.

The International Nuclear and Radiological Event Scale (INES) was introduced as a worldwide tool for promptly communicating to the public in a consistent way the safety significance of reported nuclear and radiological incidents and accidents (see Figure 51.) The scale can be applied to any event associated with nuclear facilities, as well as the transport, storage, and use of radioactive material and radiation sources.

The NRC does not require its licensees to classify events or to provide off-site notifications using the INES scale. The NRC has committed to transmit to the IAEA an INES-based rating for an applicable event occurring in the U.S.to events rated at Level 2 or above, or events attracting international public interest.

Figure 51. The International Nuclear and Radiological Event Scale

INES events are classified on the scale at 7-levels. Levels 1–3 are called "incidents" and Levels 4–7 "accidents." The scale is designed so that the severity of an event is about 10 times greater for each increase in level on the scale. Events without safety significance are called "deviations" and are classified as Below Scale or at Level 0.

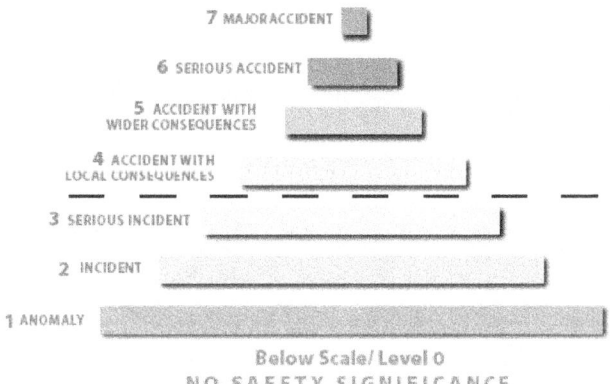

7 MAJOR ACCIDENT

6 SERIOUS ACCIDENT

5 ACCIDENT WITH WIDER CONSEQUENCES

4 ACCIDENT WITH LOCAL CONSEQUENCES

3 SERIOUS INCIDENT

2 INCIDENT

1 ANOMALY

Below Scale/ Level 0
NO SAFETY SIGNIFICANCE

Emergency Classifications

Based on NRC regulations, the emergency classifications are the sets of plant conditions that indicate various levels of risk to the public and that might require response by an offsite emergency response organization to protect citizens near the site.

Both nuclear power plants and research and test reactors use the following four emergency classifications (See Figure 52):

- **Notification of Unusual Event:** Events that indicate potential degradation in the level of safety of the plant are in progress or have occurred. No release of radioactive material requiring offsite response or monitoring is expected unless further degradation occurs.

- **Alert:** Events that involve an actual or potential substantial degradation in the level of plant safety are in progress or have occurred. Any releases of radioactive material are expected to be limited to a small fraction of the limits set forth by the U.S. Environmental Protection Agency (EPA).

- **Site Area Emergency:** Events that may result in actual or likely major failures of plant functions needed to protect the public are in progress or have occurred. Any releases of radioactive material are not expected to exceed the limits set forth by EPA except near the site boundary.

- **General Emergency:** Events that involve actual or imminent substantial core damage or melting of reactor fuel with the potential for loss of containment integrity are in progress or have occurred. Radioactive releases can be expected to exceed the limits set forth by EPA for more than the immediate site area.

Nuclear materials and fuel cycle facility licensees use the following emergency classifications:

- **Alert:** Events that could lead to a release of radioactive materials are in progress or have occurred. The release is not expected to require a response by an offsite response organization to protect citizens near the site.

- **Site Area Emergency:** Events that could lead to a significant release of radioactive materials are in progress or have occurred. The release could require a response by offsite response organizations to protect citizens near the site.

Figure 52. Emergency Classifications for Nuclear Reactor Events, 2011

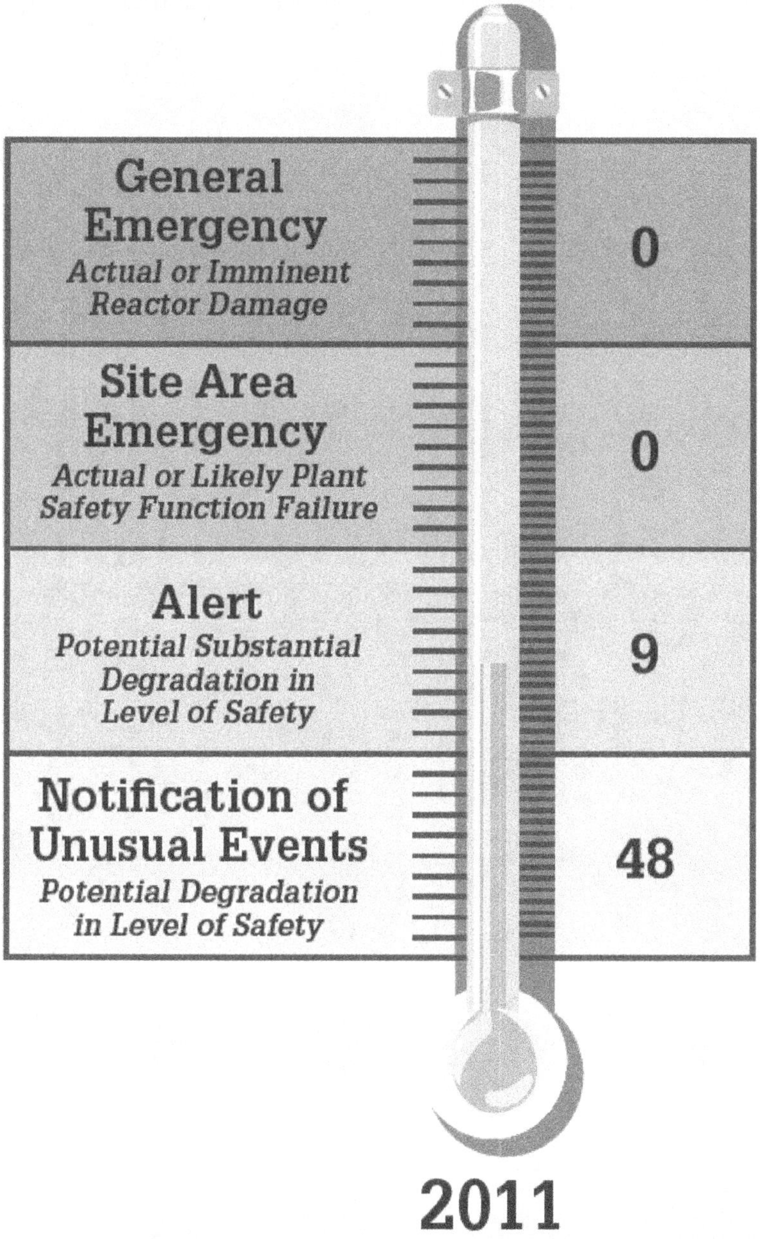

General Emergency
Actual or Imminent Reactor Damage

0

Site Area Emergency
Actual or Likely Plant Safety Function Failure

0

Alert
Potential Substantial Degradation in Level of Safety

9

Notification of Unusual Events
Potential Degradation in Level of Safety

48

2011

Appendices

Radioactive
Regulated
Waste Licensees
Abbreviations
Security
Spent Fuel
Glossary Safety
Storage Agreement
States
Research
Fuel Cycle
Materials
Nuclear Share

Agreement
Decommissioning
Enforcement
Performance
Research
Test States
Reactors
Indicators Nuclear
Licenses
Permanently
Significant Shut
Commercial Down
Power Territory
Operating

Abbreviations

ABWR	advanced boiling-water reactor
AC	Allis Chalmers
ACRS	Advisory Committee on Reactor Safeguards
AE	architect-engineer
AEC	Atomic Energy Commission (U.S.)
AEP	American Nuclear Power Company's Buchanan engineering offices
AGN	solid homogeneous core (Aerojet-General Nucleonics)
AI	Atomics International
ANS	American Nuclear Society
B&R	Burns & Roe
B&W	Babcock & Wilcox
BECH	Bechtel
BALD	Baldwin Associates
BLH	Baldwin Lima Hamilton
BRRT	Brown & Root
BWR	boiling-water reactor
CE	Combustion Engineering
CFR	*Code of Federal Regulations*
CO	Commission order
Co	company
CoC	certificate of compliance
COMM. OP.	date of commercial operation
CON TYPE	containment type
DRYAMB	dry, ambient pressure
DRYSUB	dry, subatmospheric
ICECND	wet, ice condenser
MARK 1	*wet, Mark I*
MARK 2	*wet, Mark II*
MARK 3	*wet, Mark III*
CP	construction permit
CP ISSUED	date of construction permit issuance
CVP	civil penalties
CVTR	Carolinas-Virginia Tube Reactor
CWE	Commonwealth Edison Company
CY	calendar year
DANI	Daniel International
DBDB	Duke & Bechtel
DC	design certification
DOE	Department of Energy (U.S.)
DOT	Department of Transportation (U.S.)
DUKE	Duke Power Company
EBSO	Ebasco
EIA	Energy Information Administration (DOE)
EIS	environmental impact statement
EPR	Evolutionary Power Reactor
EPZ	emergency planning zone
ERO	emergency response organization

EVESR	ESADA (Empire States Atomic Development Associates) Vallecitos Experimental Superheat Reactor
EXP. DATE	expiration date of operating license
FBR	fast breeder reactor
FLUR	Fluor Pioneer
FR	*Federal Register*
FW	Foster Wheeler
FY	fiscal year
G&H	Gibbs & Hill
GA	General Atomic
GCR	gas-cooled reactor
GEH	General Electric-Hitachi Nuclear Energy
GEIS	generic environmental impact statement
GETR	General Electric Test Reactor
GHDR	Gibbs & Hill & Durham & Richardson
GIL	Gilbert Associates
GL	general license
GPC	Georgia Power Company
GWe	gigawatt(s) electrical
HTG	high-temperature gas (reactor)
HWR	pressurized heavy-water reactor
INES	International Nuclear Event Scale
IRRS	IAEA Integrated Regulatory Review Service
ISFSI	independent spent fuel storage installation
JONES	J.A. Jones
KAIS	Kaiser Engineers
KI	potassium iodide
kW	kilowatt(s)
LES	Louisiana Energy Services
LLP	B&W lowered loop
LMFB	liquid metal fast breeder (reactor)
LR ISSUED	license renewal issued
LWGR	graphite-moderated light-water reactor
MW	megawatt(s)
MWe	megawatt(s) electrical
MWh	megawatthour(s)
MWt	megawatt(s) thermal
NIAG	Niagara Mohawk Power Corporation
NISA	Japanese Nuclear and Industrial Safety Agency
NOV	notices of violation
NOVF	notices of violation associated with inspection findings
NOVSL	notices of violation for severity level

NRC	Nuclear Regulatory Commission (U.S.)
NSP	Northern States Power Company
NSSS	nuclear steam system supplier and design type
GE 2	*GE Type 2*
GE 3	*GE Type 3*
GE 4	*GE Type 4*
GE 5	*GE Type 5*
GE 6	*GE Type 6*
WEST 2LP	*Westinghouse Two-Loop*
WEST 3LP	*Westinghouse Three-Loop*
WEST 4LP	*Westinghouse Four-Loop*
OCM	organically cooled and moderated
OL	operating license
OL ISSUED	date of latest full power operating license
PG&E	Pacific Gas & Electric Company
PHWR	pressurized heavy-water-moderated and cooled (reactor)
PRA	probabilistic risk assessment
PSE	Pioneer Services and Engineering
PSEG	Public Service Electric and Gas Company
PTHW	pressure tube heavy water
PUBS	Public Service Electric and Gas Company
PWR	pressurized-water reactor
RTR	research and test reactors
S&L	Sargent & Lundy
S&W	Stone & Webster
SCF	sodium-cooled fast (reactor)
SCGM	sodium-cooled, graphite-moderated (reactor)
SDP	significance determination process
SGEC	architect for Vogtle
SI	système internationale (d'unités) (International System of Units)
SL	site licenses
SOARCA	State-of-the-Art Consequence Analysis
SSI	Southern Services Incorporated
STARS	Strategic Teaming and Resource Sharing Group
STP	South Texas Project
TMI-2	Three Mile Island Unit 2
TRACE	reactor systems codes
TRIGA	Training Reactor and Isotopes Production, General Atomics
TVA	Tennessee Valley Authority
UE&C	United Engineers & Constructors
USAID	U.S. Agency for International Development
USEC	U.S. Enrichment Corporation
US-APWR	United States [version of] Advanced Pressurized-Water Reactor
VBWR	Vallecitos Boiling-Water Reactor
WCS	Waste Control Specialist
WDCO	Westinghouse Development Corporation
WEST	Westinghouse Electric

State and Territory Abbreviations

Alabama	AL	Kentucky	KY	Ohio	OH
Alaska	AK	Louisiana	LA	Oklahoma	OK
Arizona	AZ	Maine	ME	Oregon	OR
Arkansas	AR	Maryland	MD	Pennsylvania	PA
California	CA	Massachusetts	MA	Puerto Rico	PR
Colorado	CO	Michigan	MI	Rhode Island	RI
Connecticut	CT	Minnesota	MN	South Carolina	SC
Delaware	DE	Mississippi	MS	South Dakota	SD
District of Columbia	DC	Missouri	MO	Tennessee	TN
Florida	FL	Montana	MT	Texas	TX
Georgia	GA	Nebraska	NE	Utah	UT
Guam	GU	Nevada	NV	Vermont	VT
Hawaii	HI	New Hampshire	NH	Virgin Islands	VI
Idaho	ID	New Jersey	NJ	Virginia	VA
Illinois	IL	New Mexico	NM	Washington	WA
Indiana	IN	New York	NY	West Virginia	WV
Iowa	IA	North Carolina	NC	Wisconsin	WI
Kansas	KS	North Dakota	ND	Wyoming	WY

U.S. Commercial Nuclear Power Reactors
Operating Reactors

Plant Name, Unit Number Licensee Location Docket Number NRC Web Page Address	NRC Region	Con Type NSSS Architect Engineer Constructor	Licensed MWt	CP Issued OL Issued Comm. Op. LR Issued Exp. Date	2006–2011** Capacity Factor (Percent)
Arkansas Nuclear One, Unit 1 Entergy Operations, Inc. London, AR (6 miles NW of Russellville, AR) 050-00313 www.nrc.gov/info-finder/reactor/ano1.html	IV	PWR-DRYAMB B&W LLP BECH BECH	2,568	12/06/1968 05/21/1974 12/19/1974 06/20/2001 05/20/2034	102 94 83 99 90 87
Arkansas Nuclear One, Unit 2 Entergy Operations, Inc. London, AR (6 miles NW of Russellville, AR) 050-00368 www.nrc.gov/info-finder/reactor/ano2.html	IV	PWR-DRYAMB CE BECH BECH	3,026	12/06/1972 09/01/1978 03/26/1980 06/30/2005 07/17/2038	91 99 91 90 97 90
Beaver Valley Power Station, Unit 1 FirstEnergy Nuclear Operating Co. Shippingport, PA (17 miles W of McCandless, PA) 050-00334 www.nrc.gov/info-finder/reactor/bv1.html	I	PWR-DRYAMB WEST 3LP S&W S&W	2,900	06/26/1970 07/02/1976 10/01/1976 11/05/2009 01/29/2036	78 95 101 92 91 101
Beaver Valley Power Station, Unit 2 FirstEnergy Nuclear Operating Co. Shippingport, PA (17 miles W of McCandless, PA) 050-00412 www.nrc.gov/info-finder/reactor/bv2.html	I	PWR-DRYAMB WEST 3LP S&W S&W	2,900	05/03/1974 08/14/1987 11/17/1987 11/05/2009 05/27/2047	87 103 87 84 102 92
Braidwood Station, Unit 1 Exelon Generation Co., LLC Braceville, IL (20 miles SW of Joilet, IL) 050-00456 www.nrc.gov/info-finder/reactor/brai1.html	III	PWR-DRYAMB WEST 4LP S&L CWE	3,586.6	12/31/1975 07/02/1987 07/29/1988 N/A 10/17/2026	96 92 101 95 89 101
Braidwood Station, Unit 2 Exelon Generation Co., LLC Braceville, IL (20 miles SW of Joilet, IL) 050-00457 www.nrc.gov/info-finder/reactor/brai2.html	III	PWR-DRYAMB WEST 4LP S&L CWE	3,586.6	12/31/1975 05/20/1988 10/17/1988 N/A 12/18/2027	95 100 92 93 99 93
Browns Ferry Nuclear Plant, Unit 1 Tennessee Valley Authority Limestone County, AL (10 miles S of Athens, AL) 050-00259 www.nrc.gov/info-finder/reactor/bf1.html	II	BWR-MARK 1 GE 4 TVA TVA	3,458	05/10/1967 12/20/1973 08/01/1974 05/04/2006 12/20/2033	– 49 88 94 86 91

U.S. Commercial Nuclear Power Reactors
Operating Reactors (continued)

Plant Name, Unit Number Licensee Location Docket Number NRC Web Page Address	NRC Region	Con Type NSSS Architect Engineer Constructor	Licensed MWt	CP Issued OL Issued Comm. Op. LR Issued Exp. Date	2006– 2011** Capacity Factor (Percent)
Browns Ferry Nuclear Plant, Unit 2 Tennessee Valley Authority Limestone County, AL (10 miles S of Athens, AL) 050-00260 www.nrc.gov/info-finder/reactor/bf2.html	II	BWR-MARK 1 GE 4 TVA TVA	3,458	05/10/1967 06/28/1974 03/01/1975 05/04/2006 06/28/2034	94 78 98 94 91 80
Browns Ferry Nuclear Plant, Unit 3 Tennessee Valley Authority Limestone County, AL (10 miles S of Athens, AL) 050-00296 www.nrc.gov/info-finder/reactor/bf3.html	II	BWR-MARK 1 GE 4 TVA TVA	3,458	07/31/1968 07/02/1976 03/01/1977 05/04/2006 07/02/2036	89 93 81 95 81 87
Brunswick Steam Electric Plant, Unit 1 Carolina Power & Light Co. Southport, NC (30 miles S of Wilmington, NC) 050-00325 www.nrc.gov/info-finder/reactor/bru1.html	II	BWR-MARK 1 GE 4 UE&C BRRT	2,923	02/07/1970 09/08/1976 03/18/1977 06/26/2006 09/08/2036	87 96 85 98 83 100
Brunswick Steam Electric Plant, Unit 2 Carolina Power & Light Co. Southport, NC (30 miles S of Wilmington, NC) 050-00324 www.nrc.gov/info-finder/reactor/bru2.html	II	BWR-MARK 1 GE 4 UE&C BRRT	2,923	02/07/1970 12/27/1974 11/03/1975 06/26/2006 12/27/2034	90 87 95 80 99 79
Byron Station, Unit 1 Exelon Generation Co., LLC Byron, IL (17 miles SW of Rockford, IL) 050-00454 www.nrc.gov/info-finder/reactor/byro1.html	III	PWR-DRYAMB WEST 4LP S&L CWE	3,586.6	12/31/1975 02/14/1985 09/16/1985 N/A 10/31/2024	91 98 95 94 101 88
Byron Station, Unit 2 Exelon Generation Co., LLC Byron, IL (17 miles SW of Rockford, IL) 050-00455 www.nrc.gov/info-finder/reactor/byro2.html	III	PWR-DRYAMB WEST 4LP S&L CWE	3,586.6	12/31/1975 01/30/1987 08/02/1987 N/A 11/06/2026	102 89 96 102 96 93
Callaway Plant Union Electric Co. Fulton, MO (25 miles NE of Jefferson City, MO) 050-00483 www.nrc.gov/info-finder/reactor/call.html	IV	PWR-DRYAMB WEST 4LP BECH DANI	3,565	04/16/1976 10/18/1984 12/19/1984 N/A 10/18/2024	97 90 90 98 86 90

U.S. Commercial Nuclear Power Reactors
Operating Reactors (continued)

Plant Name, Unit Number Licensee Location Docket Number NRC Web Page Address	NRC Region	Con Type NSSS Architect Engineer Constructor	Licensed MWt	CP Issued OL Issued Comm. Op. LR Issued Exp. Date	2006– 2011** Capacity Factor (Percent)
Calvert Cliffs Nuclear Power Plant, Unit 1 Calvert Cliffs Nuclear Power Plant, LLC Lusby, MD (40 miles S of Annapolis, MD) 050-00317 www.nrc.gov/info-finder/reactor/calv1.html	I	PWR-DRYAMB CE BECH BECH	2,737	07/07/1969 07/31/1974 05/08/1975 03/23/2000 07/31/2034	84 99 93 98 90 101
Calvert Cliffs Nuclear Power Plant, Unit 2 Calvert Cliffs Nuclear Power Plant, LLC Lusby, MD (40 miles S of Annapolis, MD) 050-00318 www.nrc.gov/info-finder/reactor/calv2.html	I	PWR-DRYAMB CE BECH BECH	2,737	07/07/1969 08/13/1976 04/01/1977 03/23/2000 08/13/2036	98 90 99 93 97 92
Catawba Nuclear Station, Unit 1 Duke Energy Carolinas, LLC York, SC (18 miles S of Charlotte, NC) 050-00413 www.nrc.gov/info-finder/reactor/cat1.html	II	PWR-ICECND WEST 4LP DUKE DUKE	3,411	08/07/1975 01/17/1985 06/29/1985 12/05/2003 12/05/2043	82 102 89 91 100 89
Catawba Nuclear Station, Unit 2 Duke Energy Carolinas, LLC York, SC (18 miles S of Charlotte, NC) 050-00414 www.nrc.gov/info-finder/reactor/cat2.html	II	PWR-ICECND WEST 4LP DUKE DUKE	3,411	08/07/1975 05/15/1986 08/19/1986 12/05/2003 12/05/2043	89 84 103 90 92 101
Clinton Power Station, Unit 1 Exelon Generation Co., LLC Clinton, IL (23 miles SSE of Bloomington, IL) 050-00461 www.nrc.gov/info-finder/reactor/clin.html	III	BWR-MARK 3 GE 6 S&L BALD	3,473	02/24/1976 04/17/1987 11/24/1987 N/A 09/29/2026	90 101 99 97 92 93
Columbia Generating Station Energy Northwest Benton County, WA (12 miles NW of Richland, WA) 050-00397 www.nrc.gov/info-finder/reactor/wash2.html	IV	BWR-MARK 2 GE 5 B&R BECH	3,486	03/19/1973 04/13/1984 12/13/1984 05/22/2012 12/20/2043	94 82 93 67 95 50
Comanche Peak Nuclear Power Plant, Unit 1 Luminant Generation Co., LLC Glen Rose, TX (40 miles SW of Fort Worth, TX) 050-00445 www.nrc.gov/info-finder/reactor/cp1.html	IV	PWR-DRYAMB WEST 4LP G&H BRRT	3,612	12/19/1974 04/17/1990 08/13/1990 N/A 02/08/2030	102 85 96 100 91 91

U.S. Commercial Nuclear Power Reactors
Operating Reactors (continued)

Plant Name, Unit Number Licensee Location Docket Number NRC Web Page Address	NRC Region	Con Type NSSS Architect Engineer Constructor	Licensed MWt	CP Issued OL Issued Comm. Op. LR Issued Exp. Date	2006– 2011** Capacity Factor (Percent)
Comanche Peak Nuclear Power Plant, Unit 2 Luminant Generation Company, LLC Glen Rose, TX (40 miles SW of Fort Worth, TX) 050-00446 www.nrc.gov/info-finder/reactor/cp2.html	IV	PWR-DRYAMB WEST 4LP BECH BRRT	3,612	12/19/1974 04/06/1993 08/03/1993 N/A 02/02/2033	95 102 95 94 104 92
Cooper Nuclear Station Nebraska Public Power District Brownville, NE (23 miles S of Nebraska City, NE) 050-00298 www.nrc.gov/info-finder/reactor/cns.html	IV	BWR-MARK 1 GE 4 B&R B&R	2,419	06/04/1968 01/18/1974 07/01/1974 11/29/2010 01/18/2034	89 100 90 72 100 86
Crystal River Nuclear Generating Plant, Unit 3 Florida Power Corp. Crystal River, FL (80 miles N of Tampa, FL) 050-00302 www.nrc.gov/info-finder/reactor/cr3.html	II	PWR-DRYAMB B&W LLP GIL JONES	2,609	09/25/1968 12/03/1976 03/13/1977 N/A 12/03/2016	95 91 95 95 0 0
Davis-Besse Nuclear Power Station, Unit 1 FirstEnergy Nuclear Operating Co. Oak Harbor, OH (21 miles ESE of Toledo, OH) 050-00346 www.nrc.gov/info-finder/reactor/davi.html	III	PWR-DRYAMB B&W LLP BECH B&W	2,817	03/24/1971 04/22/1977 07/31/1978 N/A 04/22/2017	82 99 97 99 66 81
Diablo Canyon Nuclear Power Plant, Unit 1 Pacific Gas & Electric Co. Avila Beach, CA (12 miles SW of San Luis Obispo, CA) 050-00275 www.nrc.gov/info-finder/reactor/diab1.html	IV	PWR-DRYAMB WEST 4LP PG&E PG&E	3,411	4/23/1968 11/02/1984 05/07/1985 N/A 11/02/2024	101 90 98 84 88 100
Diablo Canyon Nuclear Power Plant, Unit 2 Pacific Gas & Electric Co. Avila Beach, CA 12 miles SW of San Luis Obispo, CA) 050-00323 www.nrc.gov/info-finder/reactor/diab2.html	IV	PWR-DRYAMB WEST 4LP PG&E PG&E	3,411	12/09/1970 08/26/1985 03/13/1986 N/A 08/26/2025	87 99 74 84 100 89
Donald C. Cook Nuclear Plant, Unit 1 Indiana Michigan Power Co. Bridgman, MI (13 miles S of Benton Harbor, MI) 050-00315 www.nrc.gov/info-finder/reactor/cook1.html	III	PWR-ICECND WEST 4LP AEP AEP	3,304	03/25/1969 10/25/1974 08/28/1975 08/30/2005 10/25/2034	81 103 64 3 88 87

U.S. Commercial Nuclear Power Reactors
Operating Reactors (continued)

Plant Name, Unit Number Licensee Location Docket Number NRC Web Page Address	NRC Region	Con Type NSSS Architect Engineer Constructor	Licensed MWt	CP Issued OL Issued Comm. Op. LR Issued Exp. Date	2006– 2011** Capacity Factor (Percent)
Donald C. Cook Nuclear Plant, Unit 2 Indiana Michigan Power Co. Bridgman, MI (13 miles S of Benton Harbor, MI) 050-00316 www.nrc.gov/info-finder/reactor/cook2.html	III	PWR-ICECND WEST 4LP AEP AEP	3,468	03/25/1969 12/23/1977 07/01/1978 08/30/2005 12/23/2037	89 86 101 87 84 104
Dresden Nuclear Power Station, Unit 2 Exelon Generation Co., LLC Morris, IL (25 miles SW of Joliet, IL) 050-00237 www.nrc.gov/info-finder/reactor/dres2.html	III	BWR-MARK 1 GE 3 S&L UE&C	2,957	01/10/1966 02/20/1991ᴬ 06/09/1970 10/28/2004 12/22/2029	96 92 98 91 102 95
Dresden Nuclear Power Station, Unit 3 Exelon Generation Co., LLC Morris, IL (25 miles SW of Joliet, IL) 050-00249 www.nrc.gov/info-finder/reactor/dres3.html	III	BWR-MARK 1 GE 3 S&L UE&C	2,957	10/14/1966 01/12/1971 11/16/1971 10/28/2004 01/12/2031	94 100 93 97 90 99
Duane Arnold Energy Center NextEra Energy Duane Arnold, LLC Palo, IA (8 miles NW of Cedar Rapids, IA) 050-00331 www.nrc.gov/info-finder/reactor/duan.html	III	BWR-MARK 1 GE 4 BECH BECH	1,912	06/22/1970 02/22/1974 02/01/1975 12/16/2010 02/21/2034	100 89 103 92 89 99
Edwin I. Hatch Nuclear Plant, Unit 1 Southern Nuclear Operating Co. Baxley, GA (20 miles S of Vidalia, GA) 050-00321 www.nrc.gov/info-finder/reactor/hat1.html	II	BWR-MARK 1 GE 4 BECH GPC	2,804	09/30/1969 10/13/1974 12/31/1975 01/15/2002 08/06/2034	84 98 84 94 85 98
Edwin I. Hatch Nuclear Plant, Unit 2 Southern Nuclear Operating Co., Inc. Baxley, GA (20 miles S of Vidalia, GA) 050-00366 www.nrc.gov/info-finder/reactor/hat2.html	II	BWR-MARK 1 GE 4 BECH GPC	2,804	12/27/1972 06/13/1978 09/05/1979 01/15/2002 06/13/2038	99 87 96 67 96 78
Fermi, Unit 2 The Detroit Edison Co. Newport, MI (25 miles NE of Toledo, OH) 050-00341 www.nrc.gov/info-finder/reactor/ferm2.html	III	BWR-MARK 1 GE 4 S&L DANI	3,430	09/26/1972 07/15/1985 01/23/1988 N/A 03/20/2025	76 85 98 75 80 94

A: AEC issued a provisional OL on 12/22/1969, allowing commercial operation. The NRC issued a full-term OL on 03/20/1991.

Plant Name, Unit Number Licensee Location Docket Number NRC Web Page Address	NRC Region	Con Type NSSS Architect Engineer Constructor	Licensed MWt	CP Issued OL Issued Comm. Op. LR Issued Exp. Date	2006– 2011** Capacity Factor (Percent)
Fort Calhoun Station, Unit 1 Omaha Public Power District Ft. Calhoun, NE (19 miles N of Omaha, NE) 050-00285 www.nrc.gov/info-finder/reactor/fcs.html	IV	PWR-DRYAMB CE GHDR GHDR	1,500	06/07/1968 08/09/1973 09/26/1973 11/04/2003 08/09/2033	74 104 83 100 102 28
Grand Gulf Nuclear Station, Unit 1 Entergy Operations, Inc. Port Gibson, MS (20 miles S of Vicksburg, MS) 050-00416 www.nrc.gov/info-finder/reactor/gg1.html	IV	BWR-MARK 3 GE 6 BECH BECH	3,898	09/04/1974 11/01/1984 07/01/1985 N/A 11/01/2024	94 84 86 100 88 94
H.B. Robinson Steam Electric Plant, Unit 2 Carolina Power & Light Co. Hartsville, SC (26 miles NW of Florence, SC) 050-00261 www.nrc.gov/info-finder/reactor/rob2.html	II	PWR-DRYAMB WEST 3LP EBSO EBSO	2,339	04/13/1967 07/31/1970 03/07/1971 04/19/2004 07/31/2030	104 92 87 104 57 100
Hope Creek Generating Station, Unit 1 PSEG Nuclear, LLC Hancocks Bridge, NJ (18 miles SE of Wilmington, DE) 050-00354 www.nrc.gov/info-finder/reactor/hope.html	I	BWR-MARK 1 GE 4 BECH BECH	3,840	11/04/1974 07/25/1986 12/20/1986 07/20/2011 04/11/2046	92 87 108 95 93 103
Indian Point Nuclear Generating, Unit 2 Entergy Nuclear Indian Point 2, LLC Buchanan, NY (24 miles N of New York City, NY) 050-00247 www.nrc.gov/info-finder/reactor/ip2.html	I	PWR-DRYAMB WEST 4LP UE&C WDCO	3,216	10/14/1966 09/28/1973 08/01/1974 N/A 09/28/2013	89 99 91 98 82 98
Indian Point Nuclear Generating, Unit 3 Entergy Nuclear Indian Point 3, LLC Buchanan, NY (24 miles N of New York City, NY) 050-00286 www.nrc.gov/info-finder/reactor/ip3.html	I	PWR-DRYAMB WEST 4LP UE&C WDCO	3,216	08/13/1969 12/12/1975 08/30/1976 N/A 12/12/2015	100 87 107 85 99 90
James A. FitzPatrick Nuclear Power Plant Entergy Nuclear FitzPatrick, LLC Scriba, NY (6 miles NE of Oswego, NY) 050-00333 www.nrc.gov/info-finder/reactor/fitz.html	I	BWR-MARK 1 GE 4 S&W S&W	2,536	05/20/1970 10/17/1974 07/28/1975 09/08/2008 10/17/2034	91 93 89 99 85 97

U.S. Commercial Nuclear Power Reactors
Operating Reactors (continued)

Plant Name, Unit Number Licensee Location Docket Number NRC Web Page Address	NRC Region	Con Type NSSS Architect Engineer Constructor	Licensed MWt	CP Issued OL Issued Comm. Op. LR Issued Exp. Date	2006–2011** Capacity Factor (Percent)
Joseph M. Farley Nuclear Plant, Unit 1 Southern Nuclear Operating Co. Columbia, AL (18 miles S of Dothan, AL) 050-00348 www.nrc.gov/info-finder/reactor/far1.html	II	PWR-DRYAMB WEST 3LP SSI DANI	2,775	08/16/1972 06/25/1977 12/01/1977 05/12/2005 06/25/2037	86 88 97 90 88 101
Joseph M. Farley Nuclear Plant, Unit 2 Southern Nuclear Operating Co. Columbia, AL (18 miles S of Dothan, AL) 050-00364 www.nrc.gov/info-finder/reactor/far2.html	II	PWR-DRYAMB WEST 3LP SSI BECH	2,775	08/16/1972 03/31/1981 07/30/1981 05/12/2005 03/31/2041	101 87 90 96 88 89
Kewaunee Power Station Dominion Energy Kewaunee, Inc. Kewaunee, WI (27 miles SE of Green Bay, WI) 050-00305 www.nrc.gov/info-finder/reactor/kewa.html	III	PWR-DRYAMB WEST 2LP PSE PSE	1,772	08/06/1968 12/21/1973 06/16/1974 02/24/2011 12/21/2033	75 95 90 93 102 93
LaSalle County Station, Unit 1 Exelon Generation Co., LLC Marseilles, IL (11 miles SE of Ottawa, IL) 050-00373 www.nrc.gov/info-finder/reactor/lasa1.html	III	BWR-MARK 2 GE 5 S&L CWE	3,546	09/10/1973 04/17/1982 01/01/1984 N/A 04/17/2022	93 99 100 99 94 101
LaSalle County Station, Unit 2 Exelon Generation Co., LLC Marseilles, IL (11 miles SE of Ottawa, IL) 050-00374 www.nrc.gov/info-finder/reactor/lasa2.html	III	BWR-MARK 2 GE 5 S&L CWE	3,546	09/10/1973 12/16/1983 10/19/1984 N/A 12/16/2023	102 95 94 93 101 96
Limerick Generating Station, Unit 1 Exelon Generation Co., LLC Limerick, PA (21 miles NW of Philadelphia, PA) 050-00352 www.nrc.gov/info-finder/reactor/lim1.html	I	BWR-MARK 2 GE 4 BECH BECH	3,515	06/19/1974 08/08/1985 02/01/1986 N/A 10/26/2024	93 101 95 101 91 96
Limerick Generating Station, Unit 2 Exelon Generation Co., LLC Limerick, PA (21 miles NW of Philadelphia, PA) 050-00353 www.nrc.gov/info-finder/reactor/lim2.html	I	BWR-MARK 2 GE 4 BECH BECH	3,515	06/19/1974 08/25/1989 01/08/1990 N/A 06/22/2029	100 91 101 94 99 90

Appendix

U.S. Commercial Nuclear Power Reactors
Operating Reactors (continued)

Plant Name, Unit Number Licensee Location Docket Number NRC Web Page Address	NRC Region	Con Type NSSS Architect Engineer Constructor	Licensed MWt	CP Issued OL Issued Comm. Op. LR Issued Exp. Date	2006– 2011** Capacity Factor (Percent)
McGuire Nuclear Station, Unit 1 Duke Energy Carolinas, LLC Huntersville, NC (17 miles N of Charlotte, NC) 050-00369 www.nrc.gov/info-finder/reactor/mcg1.html	II	PWR-ICECND WEST 4LP DUKE DUKE	3,411	02/23/1973 07/08/1981 12/01/1981 12/05/2003 06/12/2041	103 79 87 104 92 94
McGuire Nuclear Station, Unit 2 Duke Energy Carolinas, LLC Huntersville, NC (17 miles N of Charlotte, NC) 050-00370 www.nrc.gov/info-finder/reactor/mcg2.html	II	PWR-ICECND WEST 4LP DUKE DUKE	3,411	02/23/1973 05/27/1983 03/01/1984 12/05/2003 03/03/2043	87 103 90 94 104 94
Millstone Power Station, Unit 2 Dominion Nuclear Connecticut, Inc. Waterford, CT (3.2 miles SW of New London, CT) 050-00336 www.nrc.gov/info-finder/reactor/mill2.html	I	PWR-DRYAMB CE BECH BECH	2,700	12/11/1970 09/26/1975 12/26/1975 11/28/2005 07/31/2035	84 100 86 81 97 87
Millstone Power Station, Unit 3 Dominion Nuclear Connecticut, Inc. Waterford, CT (3.2 miles SW of New London, CT) 050-00423 www.nrc.gov/info-finder/reactor/mill3.html	I	PWR-DRYSUB WEST 4LP S&W S&W	3,650	08/09/1974 01/31/1986 04/23/1986 11/28/2005 11/25/2045	100 86 88 105 86 87
Monticello Nuclear Generating Plant, Unit 1 Northern States Power Company Monticello, MN (30 miles NW of Minneapolis, MN) 050-00263 www.nrc.gov/info-finder/reactor/mont.html	III	BWR-MARK GE 3 BECH BECH	1,775	06/19/1967 01/09/1981[B] 06/30/1971 11/08/2006 09/08/2030	101 84 97 83 94 69
Nine Mile Point Nuclear Station, Unit 1 Nine Mile Point Nuclear Station, LLC Scriba, NY (6 miles NE of Oswego, NY) 050-00220 www.nrc.gov/info-finder/reactor/nmp1.html	I	BWR-MARK 1 GE 2 NIAG S&W	1,850	04/12/1965 12/26/1974[C] 12/01/1969 10/31/2006 08/22/2029	98 88 98 92 97 84
Nine Mile Point Nuclear Station, Unit 2 Nine Mile Point Nuclear Station, LLC Scriba, NY (6 miles NE of Oswego, NY) 050-00410 www.nrc.gov/info-finder/reactor/nmp2.html	I	BWR-MARK 2 GE 5 S&W S&W	3,988	06/24/1974 07/02/1987 03/11/1988 10/31/2006 10/31/2046	90 92 90 99 89 95

B: AEC issued a provisional OL on 09/08/1970, allowing commercial operation. The NRC issued a full-term OL on 01/09/1981.

C: AEC issued a provisional OL on 08/22/1969, allowing commercial operation. The NRC issued a full-term OL on 12/26/1974.

U.S. Commercial Nuclear Power Reactors
Operating Reactors (continued)

Plant Name, Unit Number Licensee Location Docket Number NRC Web Page Address	NRC Region	Con Type NSSS Architect Engineer Constructor	Licensed MWt	CP Issued OL Issued Comm. Op. LR Issued Exp. Date	2006– 2011** Capacity Factor (Percent)
North Anna Power Station, Unit 1 Virginia Electric & Power Co. Louisa, VA (40 miles NW of Richmond, VA) 050-00338 www.nrc.gov/info-finder/reactor/na1.html	II	PWR-DRYSUB WEST 3LP S&W S&W	2,940	02/19/1971 04/01/1978 06/06/1978 03/20/2003 04/01/2038	88 89 101 92 86 78
North Anna Power Station, Unit 2 Virginia Electric & Power Co. Louisa, VA (40 miles NW of Richmond, VA) 050-00339 www.nrc.gov/info-finder/reactor/na2.html	II	PWR-DRYSUB WEST 3LP S&W S&W	2,940	02/19/1971 08/21/1980 12/14/1980 03/20/2003 08/21/2040	100 85 82 100 100 76
Oconee Nuclear Station, Unit 1 Duke Energy Carolinas, LLC Seneca, SC (30 miles W of Greenville, SC) 050-00269 www.nrc.gov/info-finder/reactor/oco1.html	II	PWR-DRYAMB B&W LLP DBDB DUKE	2,568	11/06/1967 02/06/1973 07/15/1973 05/23/2000 02/06/2033	79 99 84 85 100 79
Oconee Nuclear Station, Unit 2 Duke Energy Carolinas, LLC Seneca, SC (30 miles W of Greenville, SC) 050-00270 www.nrc.gov/info-finder/reactor/oco2.html	II	PWR-DRYAMB B&W LLP DBDB DUKE	2,568	11/06/1967 10/06/1973 09/09/1974 05/23/2000 10/06/2033	100 91 86 103 91 93
Oconee Nuclear Station, Unit 3 Duke Energy Carolinas, LLC Seneca, SC (30 miles W of Greenville, SC) 050-00287 www.nrc.gov/info-finder/reactor/oco3.html	II	PWR-DRYAMB B&W LLP DBDB DUKE	2,568	11/06/1967 07/19/1974 12/16/1974 05/23/2000 07/19/2034	91 87 102 94 91 103
Oyster Creek Nuclear Generating Station Exelon Generation Co., LLC Forked River, NJ (9 miles S of Toms River, NJ) 050-00219 www.nrc.gov/info-finder/reactor/oc.html	I	BWR-MARK 1 GE 2 B&R B&R	1,930	12/15/1964 07/02/1991[D] 12/23/1969 04/08/2009 04/09/2029	86 94 83 92 85 98
Palisades Nuclear Plant Entergy Nuclear Operations, Inc. Covert, MI (5 miles S of South Haven, MI) 050-00255 www.nrc.gov/info-finder/reactor/pali.html	III	PWR-DRYAMB CE BECH BECH	2,565.4	03/14/1967 03/24/1971 12/31/1971 01/17/2007 03/24/2031	98 86 99 90 92 96

D: AEC issued a provisional OL on 04/09/1969, allowing commercial operation. The NRC issued a full-term OL on 07/02/1991.

Appendix

U.S. Commercial Nuclear Power Reactors
Operating Reactors (continued)

Plant Name, Unit Number Licensee Location Docket Number NRC Web Page Address	NRC Region	Con Type NSSS Architect Engineer Constructor	Licensed MWt	CP Issued OL Issued Comm. Op. LR Issued Exp. Date	2006–2011** Capacity Factor (Percent)
Palo Verde Nuclear Generating Station, Unit 1 Arizona Public Service Company Wintersburg, AZ (50 miles W of Phoenix, AZ) 050-00528 www.nrc.gov/info-finder/reactor/palo1.html	IV	PWR-DRYAMB CE80-2L BECH BECH	3,990	05/25/1976 06/01/1985 01/28/1986 04/21/2011 06/01/2045	42 77 86 101 81 83
Palo Verde Nuclear Generating Station, Unit 2 Arizona Public Service Company Wintersburg, AZ (50 miles W of Phoenix, AZ) 050-00529 www.nrc.gov/info-finder/reactor/palo2.html	IV	PWR-DRYAMB CE80-2L BECH BECH	3,990	05/25/1976 04/24/1986 09/19/1986 04/21/2011 04/24/2046	85 95 74 83 101 91
Palo Verde Nuclear Generating Station, Unit 3 Arizona Public Service Company Wintersburg, AZ (50 miles W of Phoenix, AZ) 050-00530 www.nrc.gov/info-finder/reactor/palo3.html	IV	PWR-DRYAMB COMB CE80-2L BECH BECH	3,990	05/25/1976 11/25/1987 01/08/1988 04/21/2011 11/25/2047	86 64 97 83 89 97
Peach Bottom Atomic Power Station, Unit 2 Exelon Generation Co., LLC Delta, PA (17.9 miles S of Lancaster, PA) 050-00277 www.nrc.gov/info-finder/reactor/pb2.html	I	BWR-MARK 1 GE 4 BECH BECH	3,514	01/31/1968 10/25/1973 07/05/1974 05/07/2003 08/08/2033	93 101 89 102 92 101
Peach Bottom Atomic Power Station, Unit 3 Exelon Generation Co., LLC Delta, PA (17.9 miles S of Lancaster, PA) 050-00278 www.nrc.gov/info-finder/reactor/pb3.html	I	BWR-MARK 1 GE 4 BECH BECH	3,514	01/31/1968 07/02/1974 12/23/1974 05/07/2003 07/02/2034	102 93 99 89 100 90
Perry Nuclear Power Plant, Unit 1 FirstEnergy Nuclear Operating Co. Perry, OH (35 miles NE of Cleveland, OH) 050-00440 www.nrc.gov/info-finder/reactor/perr1.html	III	BWR-MARK 3 GE 6 GIL KAIS	3,758	05/03/1977 11/13/1986 11/18/1987 N/A 03/18/2026	97 75 98 67 98 79
Pilgrim Nuclear Power Station Entergy Nuclear Operations, Inc. Plymouth, MA (38 miles SE of Boston, MA) 050-00293 www.nrc.gov/info-finder/reactor/pilg.html	I	BWR-MARK 1 GE 3 BECH BECH	2,028	08/26/1968 06/08/1972 12/01/1972 05/29/2012 06/08/2032	97 85 97 90 99 85

U.S. Commercial Nuclear Power Reactors
Operating Reactors (continued)

Plant Name, Unit Number Licensee Location Docket Number NRC Web Page Address	NRC Region	Con Type NSSS Architect Engineer Constructor	Licensed MWt	CP Issued OL Issued Comm. Op. LR Issued Exp. Date	2006– 2011** Capacity Factor (Percent)
Point Beach Nuclear Plant, Unit 1 NextEra Energy Point Beach, LLC Two Rivers, WI (13 miles NW of Manitowoc, WI) 050-00266 www.nrc.gov/info-finder/reactor/poin1.html	III	PWR-DRYAMB WEST 2LP BECH BECH	1,800	07/19/1967 10/05/1970 12/21/1970 12/22/2005 10/05/2030	100 85 87 98 88 79
Point Beach Nuclear Plant, Unit 2 NextEra Energy Point Beach, LLC Two Rivers, WI (13 miles NW of Manitowoc, WI) 050-00301 www.nrc.gov/info-finder/reactor/poin2.html	III	PWR-DRYAMB WEST 2LP BECH BECH	1,800	07/25/1968 03/08/1973[E] 10/01/1972 12/22/2005 03/08/2033	91 99 89 84 96 67
Prairie Island Nuclear Generating Plant, Unit 1 Northern States Power Co.—Minnesota Welch, MN (28 miles SE of Minneapolis, MN) 050-00282 www.nrc.gov/info-finder/reactor/prai1.html	III	PWR-DRYAMB WEST 2LP FLUR NSP	1,677	06/25/1968 04/05/1974[F] 12/16/1973 06/27/2011 08/09/2033	85 92 84 97 96 91
Prairie Island Nuclear Generating Plant, Unit 2 Northern States Power Co.—Minnesota Welch, MN (28 miles SE of Minneapolis, MN) 050-00306 www.nrc.gov/info-finder/reactor/prai2.html	III	PWR-DRYAMB WEST 2LP FLUR NSP	1,677	06/25/1968 10/29/1974 12/21/1974 06/27/2011 10/29/2034	84 93 85 75 86 99
Quad Cities Nuclear Power Station, Unit 1 Exelon Generation Co., LLC Cordova, IL (20 miles NE of Moline, IL) 050-00254 www.nrc.gov/info-finder/reactor/quad1.html	III	BWR-MARK 1 GE 3 S&L UE&C	2,957	02/15/1967 12/14/1972 02/18/1973 10/28/2004 12/14/2032	89 92 96 82 99 92
Quad Cities Nuclear Power Station, Unit 2 Exelon Generation Co., LLC Cordova, IL (20 miles NE of Moline, IL) 050-00265 www.nrc.gov/info-finder/reactor/quad2.html	III	BWR-MARK 1 GE 3 S&L UE&C	2,957	02/15/1967 12/14/1972 03/10/1973 10/28/2004 12/14/2032	86 99 86 91 92 104
River Bend Station, Unit 1 Entergy Operations, Inc. St. Francisville, LA (24 miles NW of Baton Rouge, LA) 050-00458 www.nrc.gov/info-finder/reactor/rbs1.html	IV	BWR-MARK 3 GE 6 S&W S&W	3,091	03/25/1977 11/20/1985 06/16/1986 N/A 08/29/2025	88 85 82 113 98 90

E: AEC issued a provisional OL on 11/18/1971. The NRC issued a full-term OL on 03/08/1973.

F: AEC issued a provisional OL on 08/09/1973. The NRC issued a full-term OL on 04/05/1974.

U.S. Commercial Nuclear Power Reactors
Operating Reactors (continued)

Plant Name, Unit Number / Licensee / Location / Docket Number / NRC Web Page Address	NRC Region	Con Type / NSSS / Architect Engineer / Constructor	Licensed MWt	CP Issued / OL Issued / Comm. Op. / LR Issued / Exp. Date	2006–2011** Capacity Factor (Percent)
R.E. Ginna Nuclear Power Plant R.E. Ginna Nuclear Power Plant, LLC Ontario, NY (20 miles NE of Rochester, NY) 050-00244 www.nrc.gov/info-finder/reactor/ginn.html	I	PWR-DRYAMB WEST 2LP GIL BECH	1,775	04/25/1966 09/19/1969 07/01/1970 05/19/2004 09/18/2029	95 113 109 91 97 84
St. Lucie Plant, Unit 1 Florida Power & Light Co. Jensen Beach, FL (10 miles SE of Ft. Pierce, FL) 050-00335 www.nrc.gov/info-finder/reactor/stl1.html	II	PWR-DRYAMB CE EBSO EBSO	2,700	07/01/1970 03/01/1976 12/21/1976 10/02/2003 03/01/2036	102 85 91 100 72 85
St. Lucie Plant, Unit 2 Florida Power & Light Co. Jensen Beach, FL (10 miles SE of Ft. Pierce, FL) 050-00389 www.nrc.gov/info-finder/reactor/stl2.html	II	PWR-DRYAMB CE EBSO EBSO	2,700	05/02/1977 06/10/1983 08/08/1983 10/02/2003 04/06/2043	82 70 99 80 100 66
Salem Nuclear Generating Station, Unit 1 PSEG Nuclear, LLC Hancocks Bridge, NJ (18 miles SE of Wilmington, DE) 050-00272 www.nrc.gov/info-finder/reactor/salm1.html	I	PWR-DRYAMB WEST 4LP PUBS UE&C	3,459	09/25/1968 12/01/1976 06/30/1977 06/30/2011 08/13/2036	99 89 91 99 85 86
Salem Nuclear Generating Station, Unit 2 PSEG Nuclear, LLC Hancocks Bridge, NJ (18 miles SE of Wilmington, DE) 050-00311 www.nrc.gov/info-finder/reactor/salm2.html	I	PWR-DRYAMB WEST 4LP PUBS UE&C	3,459	09/25/1968 05/20/1981 10/13/1981 06/30/2011 04/18/2040	92 98 83 93 98 89
San Onofre Nuclear Generating Station, Unit 2 Southern California Edison Co. San Clemente, CA (45 miles SE of Long Beach, CA) 050-00361 www.nrc.gov/info-finder/reactor/sano2.html	IV	PWR-DRYAMB CE BECH BECH	3,438	10/18/1973 02/16/1982 08/08/1983 N/A 02/16/2022	72 89 91 60 75 105
San Onofre Nuclear Generating Station, Unit 3 Southern California Edison Co. San Clemente, CA (45 miles SE of Long Beach, CA) 050-00362 www.nrc.gov/info-finder/reactor/sano3.html	IV	PWR-DRYAMB CE BECH BECH	3,438	10/18/1973 11/15/1982 04/01/1984 N/A 11/15/2022	72 94 69 104 72 88

U.S. Commercial Nuclear Power Reactors
Operating Reactors (continued)

Plant Name, Unit Number Licensee Location Docket Number NRC Web Page Address	NRC Region	Con Type NSSS Architect Engineer Constructor	Licensed MWt	CP Issued OL Issued Comm. Op. LR Issued Exp. Date	2006– 2011** Capacity Factor (Percent)
Seabrook Station, Unit 1 NextEra Energy Seabrook, LLC Seabrook, NH (13 miles S of Portsmouth, NH) 050-00443 www.nrc.gov/info-finder/reactor/seab1.html	I	PWR-DRYAMB WEST 4LP UE&C UE&C	3,648	07/07/1976 03/15/1990 08/19/1990 N/A 03/15/2030	86 99 89 81 100 77
Sequoyah Nuclear Plant, Unit 1 Tennessee Valley Authority Soddy-Daisy, TN (16 miles NE of Chattanooga, TN) 050-00327 www.nrc.gov/info-finder/reactor/seq1.html	II	PWR-ICECND WEST 4LP TVA TVA	3,455	05/27/1970 09/17/1980 07/01/1981 N/A 09/17/2020	90 87 101 89 84 98
Sequoyah Nuclear Plant, Unit 2 Tennessee Valley Authority Soddy-Daisy, TN (16 miles NE of Chattanooga, TN) 050-00328 www.nrc.gov/info-finder/reactor/seq2.html	II	PWR-ICECND WEST 4LP TVA TVA	3,455	05/27/1970 09/15/1981 06/01/1982 N/A 09/15/2021	90 100 89 89 97 89
Shearon Harris Nuclear Power Plant, Unit 1 Carolina Power & Light Co. New Hill, NC (20 miles SW of Raleigh, NC) 050-00400 www.nrc.gov/info-finder/reactor/har1.html	II	PWR-DRYAMB WEST 3LP EBSO DANI	2,900	01/27/1978 10/24/1986 05/02/1987 12/17/2008 10/24/2046	89 94 99 94 90 103
South Texas Project, Unit 1 STP Nuclear Operating Co. Bay City, TX (90 miles SW of Houston, TX) 050-00498 www.nrc.gov/info-finder/reactor/stp1.html	IV	PWR-DRYAMB WEST 4LP BECH EBSO	3,853	12/22/1975 03/22/1988 08/25/1988 N/A 08/20/2027	91 105 95 90 101 94
South Texas Project, Unit 2 STP Nuclear Operating Co. Bay City, TX (90 miles SW of Houston, TX) 050-00499 www.nrc.gov/info-finder/reactor/stp2.html	IV	PWR-DRYAMB WEST 4LP BECH EBSO	3,853	12/22/1975 03/28/1989 06/19/1989 N/A 12/15/2028	100 93 95 101 88 88
Surry Power Station, Unit 1 Virginia Electric and Power Co. Surry, VA (17 miles NW of Newport News, VA) 050-00280 www.nrc.gov/info-finder/reactor/sur1.html	II	PWR-DRYSUB WEST 3LP S&W S&W	2,857	06/25/1968 05/25/1972 12/22/1972 03/20/2003 05/25/2032	90 89 98 94 89 101
Surry Power Station, Unit 2 Virginia Electric and Power Co. Surry, VA (17 miles NW of Newport News, VA) 050-00281 www.nrc.gov/info-finder/reactor/sur2.html	II	PWR-DRYSUB WEST 3LP S&W S&W	2,857	06/25/1968 01/29/1973 05/01/1973 03/20/2003 01/29/2033	88 101 94 92 100 76

U.S. Commercial Nuclear Power Reactors
Operating Reactors (continued)

Plant Name, Unit Number Licensee Location Docket Number NRC Web Page Address	NRC Region	Con Type NSSS Architect Engineer Constructor	Licensed MWt	CP Issued OL Issued Comm. Op. LR Issued Exp. Date	2006– 2011** Capacity Factor (Percent)
Susquehanna Steam Electric Station, Unit 1 PPL Susquehanna, LLC Berwick, Luzerne County, PA (70 miles NE of Harrisburg, PA) 050-00387 www.nrc.gov/info-finder/reactor/susq1.html	I	BWR-MARK 2 GE 4 BECH BECH	3,952	11/03/1973 07/17/1982 06/08/1983 11/24/2009 07/17/2042	86 95 89 101 80 86
Susquehanna Steam Electric Station, Unit 2 PPL Susquehanna, LLC Berwick, Luzerne County, PA (70 miles NE of Harrisburg, PA) 050-00388 www.nrc.gov/info-finder/reactor/susq2.html	I	BWR-MARK 2 GE 4 BECH BECH	3,952	11/03/1973 03/23/1984 02/12/1985 11/24/2009 03/23/2044	93 88 100 90 96 72
Three Mile Island Nuclear Station, Unit 1 Exelon Generation Co., LLC Middletown, PA (10 miles SE of Harrisburg, PA) 050-00289 www.nrc.gov/info-finder/reactor/tmi1.html	I	PWR-DRYAMB B&W LLP GIL UE&C	2,568	05/18/1968 04/19/1974 09/02/1974 10/22/2009 04/19/2034	105 97 107 86 94 92
Turkey Point Nuclear Generating, Unit 3 Florida Power & Light Co. Homestead, FL (20 miles S of Miami, FL) 050-00250 www.nrc.gov/info-finder/reactor/tp3.html	II	PWR-DRYAMB WEST 3LP BECH BECH	2,300	04/27/1967 07/19/1972 12/14/1972 06/06/2002 07/19/2032	92 97 101 86 88 96
Turkey Point Nuclear Generating, Unit 4 Florida Power & Light Co. Homestead, FL (20 miles S of Miami, FL) 050-00251 www.nrc.gov/info-finder/reactor/tp4.html	II	PWR-DRYAMB WEST 3LP BECH BECH	2,300	04/27/1967 04/10/1973 09/07/1973 06/06/2002 04/10/2033	100 86 89 99 98 84
Vermont Yankee Nuclear Power Station Entergy Nuclear Operations, Inc. Vernon, VT (5 miles S of Brattleboro, VT) 050-00271 www.nrc.gov/info-finder/reactor/vy.html	I	BWR-MARK 1 GE 4 EBSO EBSO	1,912	12/11/1967 03/21/1972 11/30/1972 03/21/2011 03/21/2032	115 87 89 99 88 90
Virgil C. Summer Nuclear Station, Unit 1 South Carolina Electric & Gas Co. Jenkinsville, SC (26 miles NW of Columbia, SC) 050-00395 www.nrc.gov/info-finder/reactor/sum.html	II	PWR-DRYAMB WEST 3LP GIL DANI	2,900	03/21/1973 11/12/1982 01/01/1984 04/23/2004 08/06/2042	89 85 87 81 100 88
Vogtle Electric Generating Plant, Unit 1 Southern Nuclear Operating Co., Inc. Waynesboro, GA (26 miles SE of Augusta, GA) 050-00424 www.nrc.gov/info-finder/reactor/vog1.html	II	PWR-DRYAMB WEST 4LP SGEC GPC	3,625.6	06/28/1974 03/16/1987 06/01/1987 06/03/2009 01/16/2047	86 99 93 91 102 92

U.S. Commercial Nuclear Power Reactors
Operating Reactors (continued)

Plant Name, Unit Number Licensee Location Docket Number NRC Web Page Address	NRC Region	Con Type NSSS Architect Engineer Constructor	Licensed MWt	CP Issued OL Issued Comm. Op. LR Issued Exp. Date	2006– 2011** Capacity Factor (Percent)
Vogtle Electric Generating Plant, Unit 2 Southern Nuclear Operating Co., Inc. Waynesboro, GA (26 miles SE of Augusta, GA) 050-00425 www.nrc.gov/info-finder/reactor/vog2.html	II	PWR-DRYAMB WEST 4LP SBEC GPC	3,625.6	06/28/1974 03/31/1989 05/20/1989 06/03/2009 02/09/2049	92 83 88 101 93 94
Waterford Steam Electric Station, Unit 3 Entergy Operations, Inc. Killona, LA (25 miles W of New Orleans, LA) 050-00382 www.nrc.gov/info-finder/reactor/wat3.html	IV	PWR-DRYAMB COMB CE EBSO EBSO	3,716	11/14/1974 03/16/1985 09/24/1985 N/A 12/18/2024	92 98 89 87 100 82
Watts Bar Nuclear Plant, Unit 1 Tennessee Valley Authority Spring City, TN (60 miles SW of Knoxville, TN) 050-00390 www.nrc.gov/info-finder/reactor/wb1.html	II	PWR-ICECND WEST 4LP TVA TVA	3,459	01/23/1973 02/07/1996 05/27/1996 N/A 11/09/2035	68 102 82 94 99 84
Wolf Creek Generating Station, Unit 1 Wolf Creek Nuclear Operating Corp. Burlington, Coffey County, KS (28 miles SE of Emporia, KS) 050-00482 www.nrc.gov/info-finder/reactor/wc.html	IV	PWR-DRYAMB WEST 4LP BECH DANI	3,565	05/31/1977 06/04/1985 09/03/1985 11/20/2008 03/11/2045	92 102 83 86 86 72

Appendix

U.S. Commercial Nuclear Power Reactors
Operating Reactors (continued)
Under Active Construction or Deferred Policy

Plant Name, Unit Number Licensee Location Docket Number NRC Web Page Address	NRC Region	Con Type NSSS Architect Engineer Constructor	Licensed MWt	CP Issued OL Issued Comm. Op. LR Issued Exp. Date	2006–2011** Capacity Factor (Percent)
Bellefonte Nuclear Power Station, Unit 1*** Tennessee Valley Authority (6 miles NE of Scottsboro, AL) 050-00438	II	PWR-DRYAMB B&W 205 TVA TVA	3,763	12/24/1974	N/A
Bellefonte Nuclear Power Station, Unit 2*** Tennessee Valley Authority (6 miles NE of Scottsboro, AL) 050-00439	II	PWR-DRYAMB B&W 205 TVA TVA	3,763	12/24/1974	N/A
Watts Bar Nuclear Plant, Unit 2**** Tennessee Valley Authority Spring City, TN (60 miles SW of Knoxville, TN) 050-00391	II	PWR-ICECND WEST 4LP TVA TVA	3,411	01/23/1973	
Virgil C. Summer Nuclear Station, Unit 2 South Carolina Electric & Gas Co. South Carolina Public Service Auth. Jenkinsville (Fairfield County), SC (26 miles NW of Columbia, SC) NPF-93	II	PWR AP1000 WEST SHAW	3,400	03/30/2012	N/A
Virgil C. Summer Nuclear Station, Unit 3 South Carolina Electric & Gas Co. South Carolina Public Service Auth. Jenkinsville (Fairfield County), SC (26 miles NW of Columbia, SC) NPF-94	II	PWR AP1000 WEST SHAW	3,400	03/30/2012	N/A
Vogtle Electric Generating Plant, Unit 3 Southern Nuclear Operating Co., Inc. Waynesboro (Burke County), GA (26 miles SE of Augusta, GA) NPF-91	II	PWR AP1000 WEST SHAW	3,400	02/10/2012	N/A
Vogtle Electric Generating Plant, Unit 4 Southern Nuclear Operating Co., Inc. Waynesboro, (Burke County), GA (26 miles SE of Augusta, GA) NPF-92	II	PWR AP1000 WEST SHAW	3,400	02/10/2012	N/A

* Note: Plant names are as identified on the license as of July 31, 2012.

** Average capacity factor is listed in year order starting with 2005.

***Bellefonte Units 1 and 2 are under the Commission Policy Statement on Deferred Plants (52 FR 38077; October 14, 1987).

****Watts Bar Unit 2 is currently under active construction.

Source: NRC, with some data compiled from EIA/DOE

U.S. Commercial Nuclear Power Reactors
Permanently Shut Down — Formerly Licensed To Operate

Unit Location	Reactor Type MWt	NSSS Vendor	OL Issued Shut Down	Decommissioning Alternative Selected Current Status
Big Rock Point Charlevoix, MI	BWR 240	GE	05/01/1964 08/29/1997	DECON DECON Completed
GE Bonus* Punta Higuera, PR	BWR 50	CE	04/02/1964 06/01/1968	ENTOMB ENTOMB
CVTR** Parr, SC	PTHW 65	WEST	11/27/1962 01/01/1967	SAFSTOR SAFSTOR
Dresden 1 Morris, IL	BWR 700	GE	09/28/1959 10/31/1978	SAFSTOR SAFSTOR
Elk River* Elk River, MN	BWR 58	AC/S&L	11/06/1962 02/01/1968	DECON DECON Completed
Fermi 1 Newport, MI	SCF 200	CE	05/10/1963 09/22/1972	DECON DECON in Progress
Fort St. Vrain Platteville, CO	HTG 842	GA	12/21/1973 08/18/1989	DECON DECON Completed
GE VBWR Sunol, CA	BWR 50	GE	08/31/1957 12/09/1963	SAFSTOR SAFSTOR
Haddam Neck Meriden, CT	PWR 1,825	WEST	12/27/1974 12/05/1996	DECON DECON Completed
Hallam* Hallam, NE	SCGM 256	BLH	01/02/1962 09/01/1964	ENTOMB ENTOMB
NS Savannah Baltimore, MD	PWR 74	B&W	08/1965 11/1970	SAFSTOR SAFSTOR
Humboldt Bay 3 Eureka, CA	BWR 200	GE	08/28/1962 07/02/1976	DECON DECON In Progress
Indian Point 1 Buchanan, NY	PWR 615	B&W	03/26/1962 10/31/1974	SAFSTOR SAFSTOR
La Crosse Genoa, WI	BWR 165	AC	07/03/1967 04/30/1987	SAFSTOR SAFSTOR
Maine Yankee Wiscasset, ME	PWR 2,700	CE	06/29/1973 12/06/1996	DECON DECON Completed
Millstone 1 Waterford, CT	BWR 2,011	GE	10/31/1970 07/21/1998	SAFSTOR SAFSTOR
Pathfinder Sioux Falls, SD	BWR 190	AC	03/12/1964 09/16/1967	DECON DECON Completed
Peach Bottom 1 Delta, PA	HTG 115	GA	01/24/1966 10/31/1974	SAFSTOR SAFSTOR

U.S. Commercial Nuclear Power Reactors
Permanently Shut Down—Formerly Licensed To Operate (continued)

Unit Location	Reactor Type MWt	NSSS Vendor	OL Issued Shut Down	Decommissioning Alternative Selected Current Status
Piqua* Piqua, OH	OCM 46	AI	08/23/1962 01/01/1966	ENTOMB ENTOMB
Rancho Seco*** Herald, CA	PWR 2,772	B&W	08/16/1974 06/07/1989	DECON DECON Completed
San Onofre 1**** San Clemente, CA	PWR 1,347	WEST	03/27/1967 11/30/1992	DECON DECON In Progress
Saxton Saxton, PA	PWR 23.5	WEST	11/15/1961 05/01/1972	DECON DECON Completed
Shippingport* Shippingport, PA	PWR 236	WEST	N/A 1982	DECON DECON Completed
Shoreham Wading River, NY	BWR 2,436	GE	04/21/1989 06/28/1989	DECON DECON Completed
Three Mile Island 2 Middletown, PA	PWR 2,770	B&W	02/08/1978 03/28/1979	(1)
Trojan Rainier, OR	PWR 3,411	WEST	11/21/1975 11/09/1992	DECON DECON Completed
Yankee-Rowe Rowe, MA	PWR 600	WEST	12/24/1963 10/01/1991	DECON DECON Completed
Zion 1 Zion, IL	PWR 3,250	WEST	10/19/1973 02/21/1997	DECON DECON In Progress
Zion 2 Zion, IL	PWR 3,250	WEST	11/14/1973 09/19/1996	DECON DECON In Progress

* AEC/DOE owned; not regulated by the NRC.

** Holds byproduct license from the State of South Carolina.

*** Low-Level radiation waste storage remains licensed by the NRC.

**** Site has been decommissioned with exception of reactor vessel in long-term storage.

Notes: See Glossary for definitions of decommissioning alternatives (DECON, ENTOMB, SAFSTOR).

(1) Three Mile Island Unit 2 has been placed in a postdefueling monitored storage mode until Unit 1 permanently ceases operation, at which time both units are planned to be decommissioned.

Source: DOE Integrated Database for 1990, "U.S. Spent Fuel and Radioactive Waste, Inventories, Projections, and Characteristics" (DOE/RW-0006, Rev. 6), and NRC, "Nuclear Power Plants in the World," Edition 6

Cancelled U.S. Commercial Nuclear Power Reactors

Unit Utility Location	Con Type MWe per Unit	Cancelled Date Status
Allens Creek 1 Houston Lighting & Power Company 4 miles NW of Wallis, TX	BWR 1,150	1982 Under CP Review
Allens Creek 2 Houston Lighting & Power Company 4 miles NW of Wallis, TX	BWR 1,150	1976 Under CP Review
Atlantic 1 & 2 Public Service Electric & Gas Company Floating Plants off the Coast of NJ	PWR 1,150	1978 Under CP Review
Bailly 1 Northern Indiana Public Service Company 12 miles NNE of Gary, IN	BWR 645	1981 With CP
Barton 1 & 2 Alabama Power & Light 15 miles SE of Clanton, AL	BWR 1,159	1977 Under CP Review
Barton 3 & 4 Alabama Power & Light 15 miles SE of Clanton, AL	BWR 1,159	1975 Under CP Review
Black Fox 1 & 2 Public Service Company of Oklahoma 3.5 miles S of Inola, OK	BWR 1,150	1982 Under CP Review
Blue Hills 1 & 2 Gulf States Utilities Company SW tip of Toledo Bend Reservoir, TX	PWR 918	1978 Under CP Review
Callaway 2 Union Electric Company 25 miles ENE of Jefferson City, MO	PWR 1,150	1981 With CP
Cherokee 1 Duke Power Company 6 miles SSW of Blacksburg, SC	PWR 1,280	1983 With CP
Cherokee 2 & 3 Duke Power Company 6 miles SSW of Blacksburg, SC	PWR 1,280	1982 With CP
Clinch River Project Management Corp., DOE, TVA 23 miles W of Knoxville, in Oak Ridge, TN	LMFB 350	1983 Under CP Review

Appendix

Cancelled U.S. Commercial Nuclear Power Reactors (continued)

Unit Utility Location	Con Type MWe per Unit	Cancelled Date Status
Clinton 2 Illinois Power Company 6 miles E of Clinton, IL	BWR 933	1983 With CP
Davis-Besse 2 & 3 Toledo Edison Company 21 miles ESE of Toledo, OH	PWR 906	1981 Under CP Review
Douglas Point 1 & 2 Potomac Electric Power Company Charles County, MD	BWR 1,146	1977 Under CP Review
Erie 1 & 2 Ohio Edison Company Berlin, OH	PWR 1,260	1980 Under CP Review
Forked River 1 Jersey Central Power & Light Company 2 miles S of Forked River, NJ	PWR 1,070	1980 With CP
Fort Calhoun 2 Omaha Public Power District 19 miles N of Omaha, NE	PWR 1,136	1977 Under CP Review
Fulton 1 & 2 Philadelphia Electric Company 17 miles S of Lancaster, PA	HTG 1,160	1975 Under CP Review
Grand Gulf 2 Entergy Nuclear Operations, Inc. 20 miles SW of Vicksburg, MS	BWR 1,250	1990 With CP
Greene County Power Authority of the State of NY 20 miles N of Kingston, NY	PWR 1,191	1980 Under CP Review
Greenwood 2 & 3 Detroit Edison Company Greenwood Township, MI	PWR 1,200	1980 Under CP Review
Hartsville A1 & A2 Tennessee Valley Authority 5 miles SE of Hartsville, TN	BWR 1,233	1984 With CP
Hartsville B1 & B2 Tennessee Valley Authority 5 miles SE of Hartsville, TN	BWR 1,233	1982 With CP

Cancelled U.S. Commercial Nuclear Power Reactors (continued)

Unit Utility Location	Con Type MWe per Unit	Cancelled Date Status
Haven 1 (formerly Koshkonong) Wisconsin Electric Power Company 4.2 miles SSW of Fort Atkinson, WI	PWR 900	1980 Under CP Review
Haven 2 (formerly Koshkonong) Wisconsin Electric Power Company 4.2 miles SSW of Fort Atkinson, WI	PWR 900	1978 Under CP Review
Hope Creek 2 Public Service Electric & Gas Company 18 miles SE of Wilmington, DE	BWR 1,067	1981 With CP
Jamesport 1 & 2 Long Island Lighting Company 65 miles E of New York City, NY	PWR 1,150	1980 With CP
Marble Hill 1 & 2 Public Service of Indiana 6 miles NE of New Washington, IN	PWR 1,130	1985 With CP
Midland 1 Consumers Power Company S of City of Midland, MI	PWR 492	1986 With CP
Midland 2 Consumers Power Company S of City of Midland, MI	PWR 818	1986 With CP
Montague 1 & 2 Northeast Nuclear Energy Company 1.2 miles SSE of Turners Falls, MA	BWR 1,150	1980 Under CP Review
New England 1 & 2 New England Power Company 8.5 miles E of Westerly, RI	PWR 1,194	1979 Under CP Review
New Haven 1 & 2 New York State Electric & Gas Corporation 3 miles NW of New Haven, NY	PWR 1,250	1980 Under CP Review
North Anna 3 Virginia Electric & Power Company 40 miles NW of Richmond, VA	PWR 907	1982 With CP
North Anna 4 Virginia Electric & Power Company 40 miles NW of Richmond, VA	PWR 907	1980 With CP

Unit Utility Location	Con Type MWe per Unit	Cancelled Date Status
North Coast 1 Puerto Rico Water Resources Authority 4.7 miles ESE of Salinas, PR	PWR 583	1978 Under CP Review
Palo Verde 4 & 5 Arizona Public Service Company 36 miles W of Phoenix, AZ	PWR 1,270	1979 Under CP Review
Pebble Springs 1 & 2 Portland General Electric Company 55 miles WSW of Tri Cities (Kenewick-Pasco-Richland, WA), OR	PWR 1,260	1982 Under CP Review
Perkins 1, 2, & 3 Duke Power Company 10 miles N of Salisbury, NC	PWR 1,280	1982 Under CP Review
Perry 2 Cleveland Electric Illuminating Co. 35 miles NE of Cleveland, OH	BWR 1,205	1994 Under CP Review
Phipps Bend 1 & 2 Tennessee Valley Authority 15 miles SW of Kingsport, TN	BWR 1,220	1982 With CP
Pilgrim 2 Boston Edison Company 4 miles SE of Plymouth, MA	PWR 1,180	1981 Under CP Review
Pilgrim 3 Boston Edison Company 4 miles SE of Plymouth, MA	PWR 1,180	1974 Under CP Review
Quanicassee 1 & 2 Consumers Power Company 6 miles E of Essexville, MI	PWR 1,150	1974 Under CP Review
River Bend 2 Gulf States Utilities Company 24 miles NNW of Baton Rouge, LA	BWR 934	1984 With CP
Seabrook 2 Public Service Co. of New Hampshire 13 miles S of Portsmouth, NH	PWR 1,198	1988 With CP
Shearon Harris 2 Carolina Power & Light Company 20 miles SW of Raleigh, NC	PWR 900	1983 With CP

Cancelled U.S. Commercial Nuclear Power Reactors (continued)

Unit Utility Location	Con Type MWe per Unit	Cancelled Date Status
Shearon Harris 3 & 4 Carolina Power & Light Company 20 miles SW of Raleigh, NC	PWR 900	1981 With CP
Skagit/Hanford 1 & 2 Puget Sound Power & Light Company 23 miles SE of Bellingham, WA	PWR 1,277	1983 Under CP Review
Sterling Rochester Gas & Electric Corporation 50 miles E of Rochester, NY	PWR 1,150	1980 With CP
Summit 1 & 2 Delmarva Power & Light Company 15 miles SSW of Wilmington, DE	HTG 1,200	1975 Under CP Review
Sundesert 1 & 2 San Diego Gas & Electric Company 16 miles SW of Blythe, CA	PWR 974	1978 Under CP Review
Surry 3 & 4 Virginia Electric & Power Company 17 miles NW of Newport News, VA	PWR 882	1977 With CP
Tyrone 1 Northern States Power Company 8 miles NE of Durond, WI	PWR 1,150	1981 Under CP Review
Tyrone 2 Northern States Power Company 8 miles NE of Durond, WI	PWR 1,150	1974 With CP
Vogtle 3 & 4 Georgia Power Company 26 miles SE of Augusta, GA	PWR 1,113	1974 With CP
Washington Nuclear 1 Energy Northwest 10 miles E of Aberdeen, WA	PWR 1,266	1995 With CP
Washington Nuclear 3 Energy Northwest 16 miles E of Aberdeen, WA	PWR 1,242	1995 With CP
Washington Nuclear 4 Energy Northwest 10 miles E of Aberdeen, WA	PWR 1,218	1982 With CP

Appendix

Cancelled U.S. Commercial Nuclear Power Reactors (continued)

Unit Utility Location	Con Type MWe per Unit	Cancelled Date Status
Washington Nuclear 5 Energy Northwest 16 miles E of Aberdeen, WA	PWR 1,242	1982 With CP
Yellow Creek 1 & 2 Tennessee Valley Authority 15 miles E of Corinth, MS	BWR 1,285	1984 With CP
Zimmer 1 Cincinnati Gas & Electric Company 25 miles SE of Cincinnati, OH	BWR 810	1984 With CP

Note: Cancellation is defined as public announcement of cancellation or written notification to the NRC. Only NRC-docketed applications are included. Status is the status of the application at the time of cancellation.

Source: DOE/EIA Commercial Nuclear Power 1991 (DOE/EIA-0438), Appendix E (page 105), and the NRC

Projected Electric Capacity Dependent on License Renewals

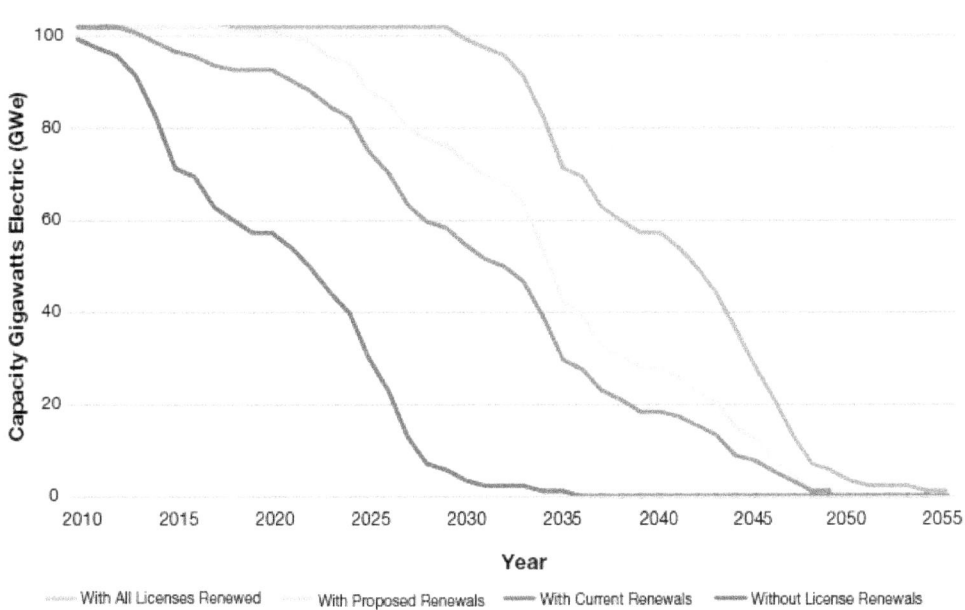

With All Licenses Renewed   With Proposed Renewals   With Current Renewals   Without License Renewals

U.S. Commercial Nuclear Power Reactors by Parent Company

Utility	NRC-Abbreviated Reactor Unit Name
AmerenUE www.ameren.com	Callaway*
Arizona Public Service Company www.aps.com	Palo Verde 1, 2, & 3*
Constellation Energy www.constellation.com	Calvert Cliffs 1 & 2 Ginna Nine Mile Point 1 & 2
Detroit Edison Company www.dteenergy.com	Fermi 2
Dominion Generation www.dom.com	Kewaunee Millstone 2 & 3 North Anna 1 & 2 Surry 1 & 2
Duke Energy Carolinas, LLC www.duke-energy.com	Catawba 1 & 2 McGuire 1 & 2 Oconee 1, 2, & 3
Energy Northwest www.energy-northwest.com	Columbia
Entergy Nuclear Operations, Inc. www.entergy-nuclear.com	Arkansas Nuclear One 1 & 2 FitzPatrick Grand Gulf 1 Indian Point 2 & 3 Palisades Pilgrim 1 River Bend 1 Vermont Yankee Waterford 3
Exelon Corporation, LLC www.exeloncorp.com	Braidwood 1 & 2 Byron 1 & 2 Clinton Dresden 2 & 3 LaSalle 1 & 2 Limerick 1 & 2 Oyster Creek Peach Bottom 2 & 3 Quad Cities 1 & 2 Three Mile Island 1
FirstEnergy Nuclear Generating Corp. www.firstenergycorp.com	Beaver Valley 1 & 2 Davis-Besse Perry 1

U.S. Commercial Nuclear Power Reactors by Parent Company (continued)

Utility	NRC-Abbreviated Reactor Unit Name
FPL Group, Inc. www.fplgroup.com	Duane Arnold Point Beach 1 & 2 Seabrook 1 St. Lucie 1 & 2 Turkey Point 3 & 4
Indiana Michigan Power Company www.indianamichiganpower.com	Cook 1 & 2
Luminant Generation Company, LLC www.luminant.com	Comanche Peak 1 & 2*
Nebraska Public Power District www.nppd.com	Cooper
Northern States Power, an Xcel Energy Operating Company www.xcelenergy.com	Monticello Prairie Island 1 & 2
Omaha Public Power District www.oppd.com	Fort Calhoun
Pacific Gas & Electric Company www.pge.com	Diablo Canyon 1 & 2*
PPL Susquehanna, LLC www.pplweb.com	Susquehanna 1 & 2
Progress Energy www.progress-energy.com	Brunswick 1 & 2 Crystal River 3 Robinson 2 Harris 1
PSEG Nuclear, LLC www.pseg.com	Hope Creek 1 Salem 1 & 2
South Carolina Electric & Gas Company www.sceg.com	Summer
Southern California Edison Company www.sce.com	San Onofre 2 & 3
Southern Nuclear Operating Company www.southerncompany.com	Hatch 1 & 2 Farley 1 & 2 Vogtle 1 & 2
STP Nuclear Operating Company www.stpnoc.com	South Texas Project 1 & 2*
Tennessee Valley Authority www.tva.gov	Browns Ferry 1, 2, & 3 Sequoyah 1 & 2 Watts Bar 1
Wolf Creek Nuclear Operating Corporation www.wcnoc.com	Wolf Creek 1*

*These plants have a joint program called the Strategic Teaming and Resource Sharing (STARS) group.
They share resources for refueling outages and develop some shared licensing applications.

APPENDIX F

U.S. Commercial Nuclear Power Reactor Operating Licenses— Issued by Year

1969 Dresden 2
Ginna
Nine Mile Point 1
Oyster Creek
1970 Point Beach 1
Robinson 2
1971 Dresden 3
Monticello
1972 Palisades
Pilgrim
Quad Cities 1
Quad Cities 2
Surry 1
Turkey Point 3
Vermont Yankee
1973 Browns Ferry 1
Fort Calhoun
Indian Point 2
Kewaunee
Oconee 1
Oconee 2
Peach Bottom 2
Point Beach 2
Surry 2
Turkey Point 4

1974 Arkansas Nuclear 1
Browns Ferry 2
Brunswick 2
Calvert Cliffs 1
Cooper
Cook 1
Duane Arnold
FitzPatrick
Hatch 1
Oconee 3
Peach Bottom 3
Prairie Island 1
Prairie Island 2
Three Mile Island 1
1975 Millstone 2
1976 Beaver Valley 1
Browns Ferry 3
Brunswick 1
Calvert Cliffs 2
Indian Point 3
Salem 1
St. Lucie 1
1977 Crystal River 3
Davis-Besse
D.C. Cook 2
Farley 1

1978 Arkansas Nuclear 2
Hatch 2
North Anna 1
1980 North Anna 2
Sequoyah 1
1981 Farley 2
McGuire 1
Salem 2
Sequoyah 2
1982 LaSalle 1
San Onofre 2
Summer
Susquehanna 1
1983 McGuire 2
San Onofre 3
St. Lucie 2
1984 Callaway
Columbia
Diablo Canyon 1
Grand Gulf 1
LaSalle 2
Susquehanna 2
1985 Byron 1
Catawba 1
Diablo Canyon 2
Fermi 2
Limerick 1

Palo Verde 1
River Bend 1
Waterford 3
Wolf Creek 1
1986 Catawba 2
Hope Creek 1
Millstone 3
Palo Verde 2
Perry 1
1987 Beaver Valley 2
Braidwood 1
Byron 2
Clinton
Harris 1
Nine Mile Point 2
Palo Verde 3
Vogtle 1
1988 Braidwood 2
South Texas Project 1
1989 Limerick 2
South Texas Project 2
Vogtle 2
1990 Comanche Peak 1
Seabrook 1
1993 Comanche Peak 2
1996 Watts Bar 1

Note: List is limited to reactors licensed to operate. Year is based on the date the initial full-power operating license was issued. NRC-abbreviated reactor names are listed.

APPENDIX G

U.S. Commercial Nuclear Power Reactor Operating Licenses— Expiration by Year, 2012–2049

2013 Indian Point 2
2015 Indian Point 3
2016 Crystal River 3
2017 Davis-Besse
2020 Sequoyah 1
2021 Sequoyah 2
2022 LaSalle 1
San Onofre 2
San Onofre 3
2023 LaSalle 2
2024 Byron 1
Callaway
Diablo Canyon 1
Grand Gulf 1
Limerick 1
Waterford 3
2025 Diablo Canyon 2
Fermi 2
River Bend 1
2026 Braidwood 1
Byron 2
Clinton
Perry
2027 Braidwood 2
South Texas Project 1

2028 South Texas Project 2
2029 Dresden 2
Ginna
Limerick 2
Nine Mile Point 1
Oyster Creek
2030 Comanche Peak 1
Monticello
Point Beach 1
Robinson 2
Seabrook
2031 Dresden 3
Palisades
2032 Quad Cities 1
Quad Cities 2
Surry 1
Turkey Point 3
Vermont Yankee
Pilgrim
2033 Browns Ferry 1
Comanche Peak 2
Fort Calhoun
Kewaunee
Oconee 1
Oconee 2
Peach Bottom 2

Point Beach 2
Prairie Island 1
Surry 2
Turkey Point 4
2034 Arkansas Nuclear 1
Browns Ferry 2
Brunswick 2
Calvert Cliffs 1
Cook 1
Cooper
Duane Arnold
Hatch 1
FitzPatrick
Oconee 3
Peach Bottom 3
Prairie Island 2
Three Mile Island 1
2035 Millstone 2
Watts Bar 1
2036 Beaver Valley 1
Browns Ferry 3
Brunswick 1
Calvert Cliffs 2
St. Lucie 1
Salem 1
2037 Cook 2
Farley 1

2038 Arkansas Nuclear 2
Hatch 2
North Anna 1
2040 North Anna 2
Salem 2
2041 Farley 2
McGuire 1
2042 Summer
Susquehanna 1
2043 Catawba 1
Catawba 2
McGuire 2
St. Lucie 2
Columbia
2044 Susquehanna 2
2045 Millstone 3
Palo Verde 1
Wolf Creek 1
2046 Nine Mile Point 2
Harris 1
Hope Creek
Palo Verde 2
2047 Beaver Valley 2
Palo Verde 3
Vogtle 1
2049 Vogtle 2

Note: Limited to reactors licensed to operate. NRC-abbreviated reactor names listed. Data are as of July 2011.

Appendix

APPENDIX H
Industry Performance Indicators:
Annual Industry Averages, FYs 2002–2011

Indicator	2002	2003	2004	2005	2006	2007	2008	2009	2010	2011
Automatic Scrams	0.44	0.75	0.56	0.47	0.32	0.48	0.29	0.36	0.44	0.45
Safety System Actuations	0.18	0.41	0.24	0.38	0.22	0.25	0.14	0.23	0.18	0.19
Significant Events	0.05	0.07	0.04	0.05	0.03	0.02	0.03	0.02	0.10	0.06
Safety System Failures	0.88	0.96	0.78	0.99	0.59	0.68	0.71	0.71	0.97	0.92
Forced Outage Rate	1.70	3.04	1.88	2.44	1.47	1.43	1.34	2.21	1.74	1.80
Equipment-Forced Outage Rate	0.12	0.16	0.15	0.13	0.10	0.11	0.08	0.09	0.10	0.09
Collective Radiation Exposure	111	125	100	117	93	110	96	87	91	91
Drill/Exercise Performance	95	96	96	96	96	98	96	97	97	97
ERO Drill Participation	97	98	98	98	98	98	98	99	99	99
Alert and Notification System Reliability	99	99	99	99	99	99	100	100	100	100

Safety System Failures

Safety system failures are any actual failures, events, or conditions that could prevent a system from performing its required safety function.

Industry Performance Indicators:
Annual Industry Averages, FYs 2002–2011 (continued)

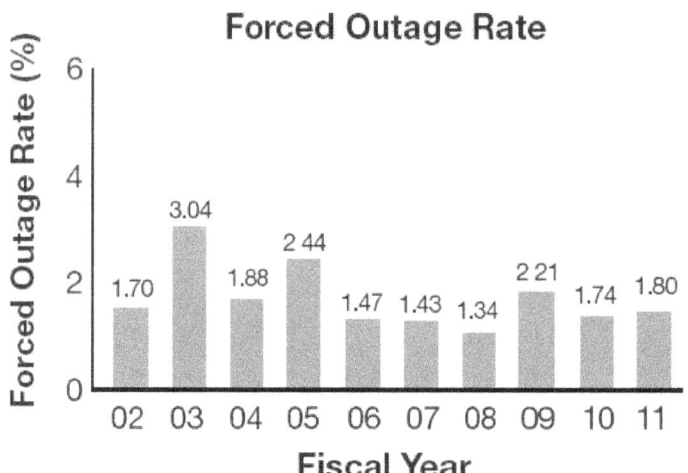

Forced Outage Rate

The forced outage rate is the number of hours that the plant is unable to operate (forced outage hours) divided by the sum of the hours that the plant is generating and transmitting electricity (unit service hours) and the hours that the plant is unable to operate (forced outage hours).

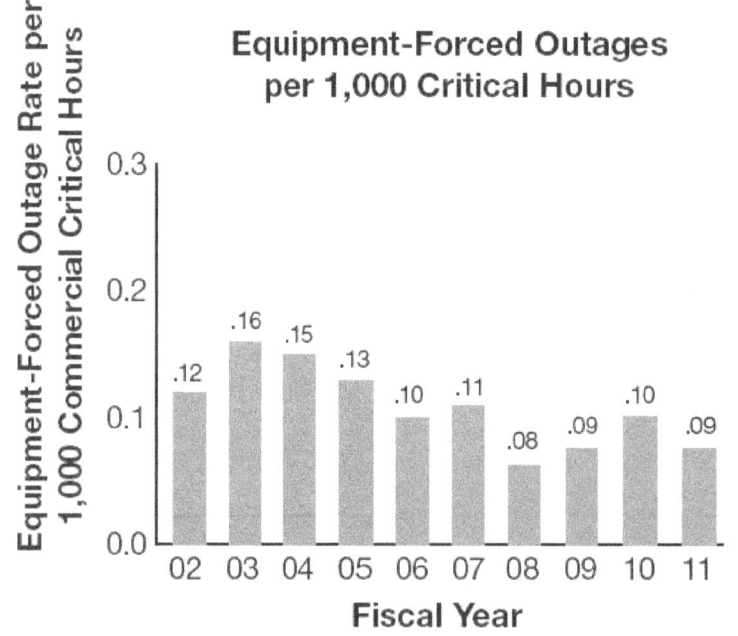

Equipment-Forced Outages per 1,000 Critical Hours

This indicator is the number of times the plant is forced to shut down because of equipment failures for every 1,000 hours that the plant is in operation and transmitting electricity.

Appendix

Industry Performance Indicators:
Annual Industry Averages, FYs 2002–2011 (continued)

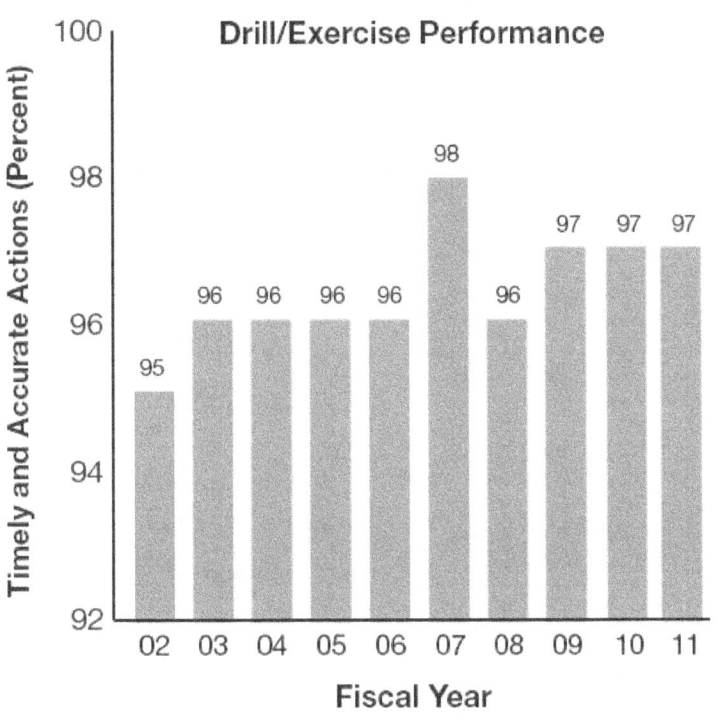

The percentage of timely and accurate actions taken by plant personnel (emergency classifications, protective action recommendations, and notification to offsite authorities) in drills and actual events during the previous 2 years.

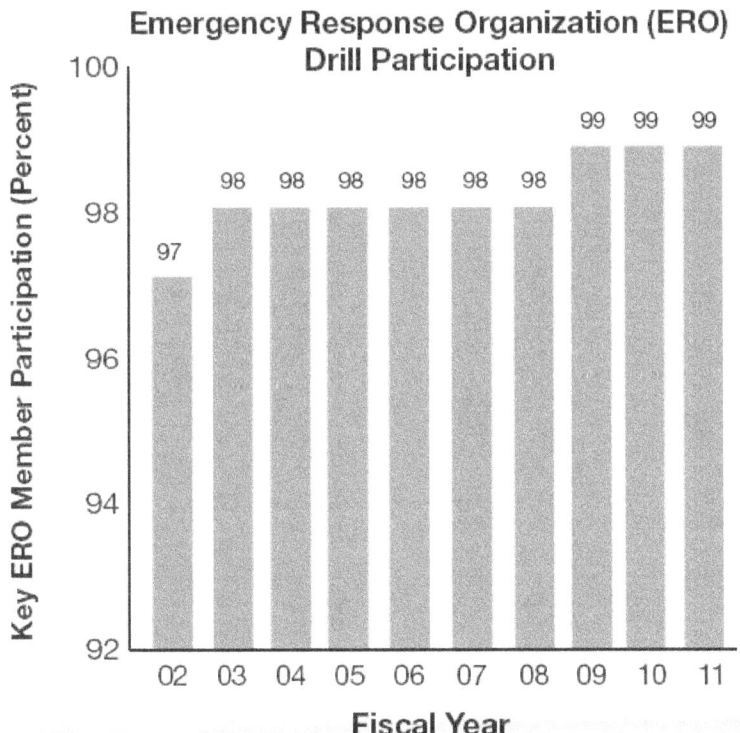

The percentage of participation by key plant personnel in drills or actual events in the previous 2 years, indicating proficiency and readiness to respond to emergencies.

Operating U.S. Nuclear Research and Test Reactors
Regulated by the NRC

Licensee Location	Reactor Type OL Issued	Power Level (kW)	Licensee Number Docket Number
Aerotest San Ramon, CA	TRIGA (Indus) 07/02/1965	250	R-98 50-228
Armed Forces Radiobiology Research Institute Bethesda, MD	TRIGA 06/26/1962	1,100	R-84 50-170
Dow Chemical Company Midland, MI	TRIGA 07/03/1967	300	R-108 50-264
GE-Hitachi Sunol, CA	Tank 10/31/1957	100	R-33 50-73
Idaho State University Pocatello, ID	AGN-201 #103 10/11/1967	0.005	R-110 50-284
Kansas State University Manhattan, KS	TRIGA 10/16/1962	250	R-88 50-188
Massachusetts Institute of Technology Cambridge, MA	HWR Reflected 06/09/1958	6,000	R-37 50-20
National Institute of Standards & Technology Gaithersburg, MD	Nuclear Test 05/21/1970	20,000	TR-5 50-184
North Carolina State University Raleigh, NC	Pulstar 08/25/1972	1,000	R-120 50-297
Ohio State University Columbus, OH	Pool 02/24/1961	500	R-75 50-150
Oregon State University Corvallis, OR	TRIGA Mark II 03/07/1967	1,100	R-106 50-243
Pennsylvania State University State College, PA	TRIGA 07/08/1955	1,100	R-2 50-5
Purdue University West Lafayette, IN	Lockheed 08/16/1962	1	R-87 50-182
Reed College Portland, OR	TRIGA Mark I 07/02/1968	250	R-112 50-288
Rensselaer Polytechnic Institute Troy, NY	Critical Assembly 07/03/1964	0.1	CX-22 50-225
Rhode Island Atomic Energy Commission Narragansett, RI	GE Pool 07/23/1964	2,000	R-95 50-193

Operating U.S. Nuclear Research and Test Reactors Regulated by the NRC (continued)

Licensee Location	Reactor Type OL Issued	Power Level (kW)	Licensee Number Docket Number
Texas A&M University College Station, TX	AGN-201M #106 08/26/1957	0.005	R-23 50-59
Texas A&M University College Station, TX	TRIGA 12/07/1961	1,000	R-128 50-128
U.S. Geological Survey Denver, CO	TRIGA Mark I 02/24/1969	1,000	R-113 50-274
University of California/Davis Sacramento, CA	TRIGA 08/13/1998	2,300	R-130 50-607
University of California/Irvine Irvine, CA	TRIGA Mark I 11/24/1969	250	R-116 50-326
University of Florida Gainesville, FL	Argonaut 05/21/1959	100	R-56 50-83
University of Maryland College Park, MD	TRIGA 10/14/1960	250	R-70 50-166
University of Massachusetts/Lowell Lowell, MA	GE Pool 12/24/1974	1,000	R-125 50-223
University of Missouri/Columbia Columbia, MO	Tank 10/11/1966	10,000	R-103 50-186
University of Missouri/Rolla Rolla, MO	Pool 11/21/1961	200	R-79 50-123
University of New Mexico Albuquerque, NM	AGN-201M #112 09/17/1966	0.005	R-102 50-252
University of Texas Austin, TX	TRIGA Mark II 01/17/1992	1,100	R-129 50-602
University of Utah Salt Lake City, UT	TRIGA Mark I 09/30/1975	100	R-126 50-407
University of Wisconsin Madison, WI	TRIGA 11/23/1960	1,000	R-74 50-156
Washington State University Pullman, WA	TRIGA 03/06/1961	1,000	R-76 50-27

U.S. Nuclear Research and Test Reactors
Under Decommissioning Regulated by the NRC

Licensee Location	Reactor Type Power Level (kW)	OL Issued Shutdown	Decommissioning Alternative Selected Current Status
General Atomics San Diego, CA	TRIGA Mark F 1,500	07/01/60 09/07/94	DECON SAFSTOR
General Atomics San Diego, CA	TRIGA Mark I 250	05/03/58 12/17/96	DECON SAFSTOR
General Electric Company Sunol, CA	GETR (Tank) 50,000	01/07/59 06/26/85	SAFSTOR SAFSTOR
General Electric Company Sunol, CA	EVESR 17,000	11/12/63 02/01/67	SAFSTOR SAFSTOR
National Aeronautics and Space Administration Sandusky, OH	Test 60,000	05/02/62 07/07/73	DECON DECON In Progress
National Aeronautics and Space Administration Sandusky, OH	Mockup 100	06/14/61 07/07/73	DECON DECON In Progress
University of Buffalo Buffalo, NY	Pulstar 2,000	03/24/61 07/23/96	DECON SAFSTOR In Progress
University of Illinois Urbana-Champaign, IL	TRIGA 1,500	07/22/69 04/12/99	SAFSTOR DECON In Progress
University of Michigan Ann Arbor, MI	Pool 2,000	09/13/57 01/29/04	DECON DECON In Progress
Veterans Administration Omaha, NE	TRIGA 20	06/26/59 11/05/01	DECON SAFSTOR
Worcester Polytechnic Institute Worcester, MA	GE 10	12/16/59 06/30/07	DECON DECON Pending

Appendix

APPENDIX K
Agreement States

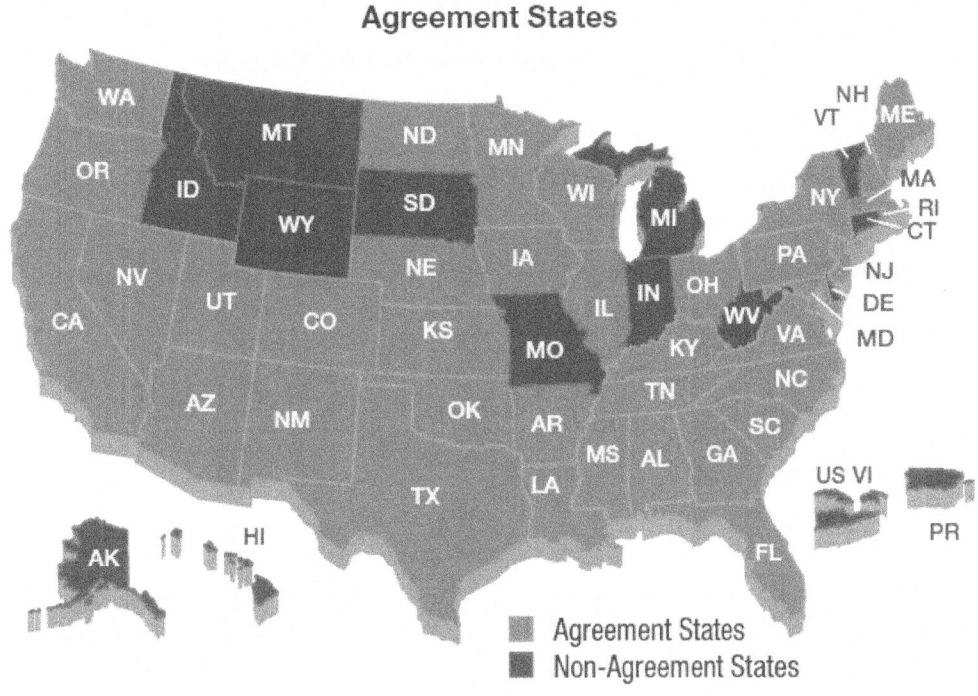

- ■ Agreement States
- ■ Non-Agreement States

APPENDIX L
State Electricity Profile by Nuclear Source

State	Net Generation	State	Net Generation	State	Net Generation
Alabama	25.87%	Kentucky	0.00%	North Dakota	0.00%
Alaska	0.00%	Louisiana	16.15%	Ohio	10.37%
Arizona	27.99%	Maine	0.00%	Oklahoma	0.00%
Arkansas	23.27%	Maryland	33.02%	Oregon	0.00%
California	17.96%	Massachusetts	11.88%	Pennsylvania	33.14%
Colorado	0.00%	Michigan	29.48%	Rhode Island	0.00%
Connecticut	47.76%	Minnesota	22.29%	South Carolina	50.79%
Delaware	0.00%	Mississippi	18.97%	South Dakota	0.00%
District of Columbia	0.00%	Missouri	10.15%	Tennessee	32.69%
Florida	9.61%	Montana	0.00%	Texas	9.63%
Georgia	23.48%	Nebraska	18.93%	Utah	0.00%
Hawaii	0.00%	Nevada	0.00%	Vermont	74.13%
Idaho	0.00%	New Hampshire	37.68%	Virginia	35.01%
Illinois	47.59%	New Jersey	51.16%	Washington	4.64%
Indiana	0.00%	New Mexico	0.00%	West Virginia	0.00%
Iowa	9.07%	New York	31.19%	Wisconsin	17.97%
Kansas	15.27%	North Carolina	31.49%	Wyoming	0.00%

Source: DOE/EIA, "State Electricity Profiles," data from May 2012, www.eia.doe.gov

Major U.S. Fuel Cycle Facility Sites

Licensee	Location	Status
Uranium Hexafluoride Conversion Facility		
Honeywell International, Inc.	Metropolis, IL	active
Uranium Fuel Fabrication Facilities		
Global Nuclear Fuels-Americas, LLC	Wilmington, NC	active
Westinghouse Electric Company, LLC Columbia Fuel Fabrication Facility	Columbia, SC	active
Nuclear Fuel Services, Inc.	Erwin, TN	active
AREVA NP, Inc. Mt. Athos Road Facility	Lynchburg, VA	inactive, license termination pending
B&W Nuclear Operations Group	Lynchburg, VA	active
AREVA NP, Inc.	Richland, WA	active
Mixed Oxide Fuel Fabrication Facility		
Shaw AREVA MOX Services, LLC	Aiken, SC	under construction (operating license under review)
Gaseous Diffusion Uranium Enrichment Facilities		
USEC Inc.	Paducah, KY	active
Gas Centrifuge Uranium Enrichment Facilities		
USEC Inc.	Piketon, OH	under construction
Louisiana Energy Services (URENCO-USA)	Eunice, NM	active*
AREVA Enrichment Services LLC Eagle Rock Enrichment Facilities	Idaho Falls, ID	active**
Laser Separation Enrichment Facility		
GE-Hitachi	Wilmington, NC	under review
Uranium Hexafluoride Deconversion Facility		
International Isotopes	Hobbs, NM	under review

* Partially operating and producing enriched uranium while undergoing further phases of construction.

** NRC issued license in Oct. 2011 and construction on the facility has not begun.

Note: The NRC regulates nine other facilities that possess significant quantities of special nuclear material (other than reactors) or process source material (other than uranium recovery facilities).

Data are as of July 2012.

Appendix

APPENDIX N
Dry Spent Fuel Storage Designs:
NRC-Approved for Use by General Licensees

Vendor	Docket #	Storage Design Model
General Nuclear Systems, Inc.	72-1000	CASTOR V/21
NAC International, Inc.	72-1002	NAC S/T
	72-1003	NAC-C28 S/T
	72-1015	NAC-UMS
	72-1025	NAC-MPC
	72-1031	Magnastor
Holtec International	72-1008	HI-STAR 100
	72-1014	HI-STORM 100
	72-1032	HI-STORM FW
Energy Solutions, Inc.	72-1007	VSC-24
	72-1026	Fuel Solutions™ (WSNF-220, -221, -223)
		W-150 Storage Cask
		W-100 Transfer Cask
		W-21, W-74 Canisters
Transnuclear, Inc.	72-1005	TN-24
	72-1027	TN-68
	72-1021	TN-32, 32A, 32B
	72-1004	Standardized NUHOMS®-24P, -24PHB, -24PTH, -32PT, -32PTH1, -52B, -61BT, -61BTH
	72-1029	Standardized Advanced NUHOMS®-24PT1, -24PT4
	72-1030	NUHOMS® HD-32PTH

Data are as of June 2012 (See latest list on the NRC Web site at www.nrc.gov/waste/spent-fuel-storage/designs.html.)

Dry Cask Spent Fuel Storage Licensees

Name Licensee	License Type	Date Issued	Vendor	Storage Model	Docket #
Surry Virginia Electric & Power Company (Dominion Gen.)	SL	07/02/1986	General Nuclear Systems, Inc. Transnuclear, Inc. General Nuclear Westinghouse, Inc.	CASTOR V/21 TN-32 NAC-128 CASTOR X/33 MC-10	72-2
	GL	08/06/2007	Transnuclear, Inc.	NUHOMS®-HD	72-55
H.B. Robinson Carolina Power & Light Company	SL GL	08/13/1986 09/06/2005	Transnuclear, Inc. Transnuclear, Inc.	NUHOMS®-7P NUHOMS®-24P	72-3 72-60
Oconee Duke Energy Company	SL GL	01/29/1990 03/05/1999	Transnuclear, Inc. Transnuclear, Inc.	NUHOMS®-24P NUHOMS®-24P	72-4 72-40
Fort St. Vrain* U.S. Department of Energy	SL	11/04/1991	FW Energy Applications, Inc.	Modular Vault Dry Store	72-9
Calvert Cliffs Calvert Cliffs Nuclear Power Plant, Inc.	SL	11/25/1992	Transnuclear, Inc.	NUHOMS®-24P NUHOMS®-32P	72-8
Palisades Entergy Nuclear Operations, Inc.	GL	05/11/1993	Energy Solutions, Inc.	VSC-24 NUHOMS®-32PT	72-7
Prairie Island Northern States Power Co., a Minnesota Corp.	SL	10/19/1993	Transnuclear, Inc.	TN-40 HT TN-40	72-10
Point Beach FLP Energy Point Beach, LLC	GL	05/26/1996	Energy Solutions, Inc.	VSC-24 NUHOMS®-32PT	72-5
Davis-Besse FirstEnergy Nuclear Operating Company	GL	01/01/1996	Transnuclear, Inc.	NUHOMS®-24P	72-14
Arkansas Nuclear Entergy Nuclear Operations, Inc.	GL	12/17/1996	Energy Solutions, Inc. Holtec International	VSC-24 HI-STORM 100	72-13
North Anna Virginia Electric & Power Company (Dominion Gen.)	SL GL	06/30/1998 03/10/2008	Transnuclear, Inc. Transnuclear, Inc.	TN-32 NUHOMS®-HD	72-16 72-56
Trojan Portland General Electric Corp.	SL	03/31/1999	Holtec International	HI-STORM 100	72-17

Appendix

Dry Cask Spent Fuel Storage Licensees (continued)

Name Licensee	License Type	Date Issued	Vendor	Storage Model	Docket #
Idaho National Lab TMI-2 Fuel Debris, U.S. Department of Energy	SL	03/19/1999	Transnuclear, Inc.	NUHOMS®-12T	72-20
Susquehanna PPL Susquehanna, LLC	GL	10/18/1999	Transnuclear, Inc.	NUHOMS®-52B NUHOMS®-61BT	72-28
Peach Bottom Exelon Generation Company, LLC	GL	06/12/2000	Transnuclear, Inc.	TN-68	72-29
Hatch Southern Nuclear Operating, Inc.	GL	07/06/2000	Holtec International	HI-STAR 100 HI-STORM 100	72-36
Dresden Exelon Generation Company, LLC	GL	07/10/2000	Holtec International	HI-STAR 100 HI-STORM 100	72-37
Rancho Seco Sacramento Municipal Utility District	SL	06/30/2000	Transnuclear, Inc.	NUHOMS®-24P	72-11
McGuire Duke Energy, LLC	GL	02/01/2001	Transnuclear, Inc.	TN-32	72-38
Big Rock Point Entergy Nuclear Operations, Inc.	GL	11/18/2002	Energy Solutions, Inc.	Fuel Solutions™ W74	72-43
James A. FitzPatrick Entergy Nuclear Operations, Inc.	GL	04/25/2002	Holtec International	HI-STORM 100	72-12
Maine Yankee Maine Yankee Atomic Power Company	GL	08/24/2002	NAC International, Inc.	NAC-UMS	72-30
Columbia Generating Station Energy Northwest	GL	09/02/2002	Holtec International	HI-STORM 100	72-35
Oyster Creek AmerGen Energy Company, LLC.	GL	04/11/2002	Transnuclear, Inc.	NUHOMS®-61BT	72-15
Yankee Rowe Yankee Atomic Electric	GL	06/26/2002	NAC International, Inc.	NAC-MPC	72-31
Duane Arnold Next Era Energy Duane Arnold, LLC.	GL	09/01/2003	Transnuclear, Inc.	NUHOMS®-61BT	72-32

Dry Cask Spent Fuel Storage Licensees (continued)

Name Licensee	License Type	Date Issued	Vendor	Storage Model	Docket #
Palo Verde Arizona Public Service Co.	GL	03/15/2003	NAC International, Inc.	NAC-UMS	72-44
San Onofre Southern California Edison Company	GL	10/03/2003	Transnuclear, Inc.	NUHOMS®-24PT	72-41
Diablo Canyon Pacific Gas & Electric Co.	SL	03/22/2004	Holtec International	HI-STORM 100	72-26
Haddam Neck CT Yankee Atomic Power	GL	05/21/2004	NAC International, Inc.	NAC-MPC	72-39
Sequoyah Tennessee Valley Authority	GL	07/13/2004	Holtec International	HI-STORM 100	72-34
Idaho Spent Fuel Facility	SL	11/30/2004	Foster Wheeler Environmental Corp.	Concrete Vault	72-25
Humboldt Bay Pacific Gas & Electric Co.	SL	11/30/2005	Holtec International	HI-STORM 100HB	72-27
Private Fuel Storage Facility	SL	02/21/2006	Holtec International	HI-STORM 100	72-22
Browns Ferry Tennessee Valley Authority	GL	08/21/2005	Holtec International	HI-STORM 100S	72-52
Joseph M. Farley Southern Nuclear Operating Co.	GL	08/25/2005	Transnuclear, Inc.	NUHOMS®-32PT	72-42
Millstone Dominion Generation	GL	02/15/2005	Transnuclear, Inc.	NUHOMS®-32PT	72-47
Quad Cities Exelon Generation Company, LLC	GL	12/02/2005	Holtec International	HI-STORM 100S	72-53
River Bend Entergy Nuclear Operations, Inc.	GL	12/29/2005	Holtec International	HI-STORM 100S	72-49
Fort Calhoun Omaha Public Power District	GL	07/29/2006	Transnuclear, Inc.	NUHOMS®-32PT	72-54
Hope Creek/Salem PSEG, Nuclear, LLC	GL	11/10/2006	Holtec International	HI-STORM 100	72-48
Grand Gulf Entergy Nuclear Operations, Inc.	GL	11/18/2006	Holtec International	HI-STORM 100S	72-50
Catawba Duke Energy Carolinas, LLC	GL	07/30/2007	NAC International, Inc.	NAC-UMS	72-45

Dry Cask Spent Fuel Storage Licensees (continued)

Name Licensee	License Type	Date Issued	Vendor	Storage Model	Docket #
Indian Point Entergy Nuclear Operations, Inc.	GL	01/11/2008	Holtec International	HI-STORM 100	72-51
St. Lucie Florida Power and Light Company	GL	03/14/2008	Transnuclear, Inc.	NUHOMS®-HD	72-61
Vermont Yankee Entergy Nuclear Operations, Inc.	GL	05/25/2008	Holtec International	HI-STORM100	72-59
Limerick Exelon Generation Co., LLC	GL	08/01/2008	Transnuclear, Inc.	NUHOMS®-61BT	72-65
Seabrook FPL Energy	GL	08/07/2008	Transnuclear, Inc.	NUHOMS®-HD-3PTM	72-61
Monticello Northern States Power Co.	GL	09/17/2008	Transnuclear, Inc.	NUHOMS®-61BT	72-58
Kewaunee Northern States Power Co.	GL	09/11/2009	Transnuclear, Inc.	NUHOMS®-39PT	72-64
Byron Exelon Generation Co., LLC	GL	09/09/2010	Holtec International	HI-STORM 100	72-68
Cooper Nuclear Station Nebraska Public Power District	GL	10/21/2010	Transnuclear, Inc.	NUHOMS-61BT	72-66
La Salle Exelon Generation Co., LLC	GL	11/01/2010	Holtec International	HI-STORM100	72-70
Turkey Point ISFSI Florida Power and Light Company	GL	07/29/2010	Transnuclear, Inc.	NUHOMS HD	72-62
Waterford Steam Electric Station Entergy Nuclear Operations, Inc.	GL	11/08/11	Holtec International	HI-STORM 100	72-75
Braidwood Exelon Generation Co., LLC	GL	11/23/11	Holtec International	HI-STORM 100	72-73
Comanche Peak Luminant Generation Company, LLC	GL	2/28/12	Holtec International	HI-STORM 100	72-74

*Fort St. Vrain is undergoing decommissioning and was transferred to DOE on June 4, 1999.

Note: NRC-abbreviated unit names.

U.S. Low-Level Radioactive Waste Compacts

Appalachian

Delaware
Maryland
Pennsylvania
West Virginia

Atlantic

Connecticut
New Jersey
South Carolina*

Central

Arkansas
Kansas
Louisiana
Oklahoma

Central Midwest

Illinois
Kentucky

Midwest

Indiana
Iowa
Minnesota
Missouri
Ohio
Wisconsin

Northwest

Alaska
Hawaii
Idaho
Montana
Oregon
Utah*
Washington*
Wyoming

Rocky Mountain

Colorado
Nevada
New Mexico
*(Northwest accepts Rocky
Mountain waste as agreed
between compacts)*

Southeast

Alabama
Florida
Georgia
Mississippi
Tennessee
Virginia

Southwestern

Arizona
California
North Dakota
South Dakota

Texas

Texas*
Vermont

Unaffiliated

District of Columbia
Maine
Massachusetts
Michigan
Nebraska
New Hampshire
New York
North Carolina
Puerto Rico
Rhode Island

Note: Data are as of June 2012.
*Site of an active LLW disposal facility.

Appendix

Storage of Commercial Spent Fuel by State through 2011

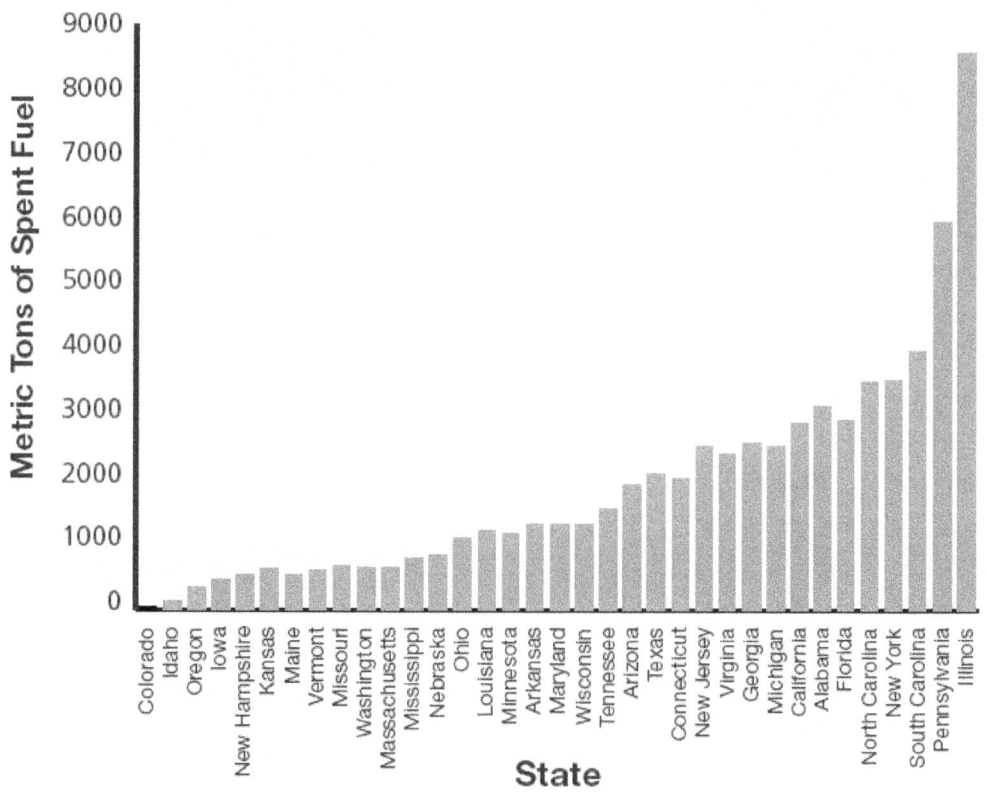

Idaho is holding used fuel from Three Mile Island 2 and the used Fuel Data are rounded up to the nearest 10 for CY 2011.

Source: Gutherman Technical Services and Department of Energy

Updated: April 12, 2012.

APPENDIX R
NRC-Regulated Complex Material Sites
Undergoing Decommissioning

Company	Location
AAR Manufacturing, Inc. (Brooks & Perkins)	Livonia, MI
ABB, Inc.	Windsor, CT
Analytical Bio-Chemistry Laboratories	Columbia, MO
Army, Department of, Jefferson Proving Ground	Madison, IN
Babcock & Wilcox SLDA	Vandergrift, PA
Beltsville Agricultural Research Center	Beltsville, MD
FMRI	Muskogee, OK
Hunter's Point Naval Shipyard	San Francisco, CA
Kerr-McGee	Cimarron, OK
Mallinckrodt Chemical, Inc.	St. Louis, MO
McClellan Air Force Base	Sacramento, CA
NWI Breckenridge	Breckenridge, MI
Pohakuloa Training Area	Kawaihe Harbor, HI
Schofield Army Barracks	Wahiawa, HI
Sigma Aldrich	Maryland Heights, MO
Stepan Chemical Corporation	Maywood, NJ
UNC Naval Products	New Haven, CT
West Valley Demonstration Project	West Valley, NY
Westinghouse Electric Corporation—Hematite	Festus, MO

Note: Data are as of June 2012.

Appendix

Nuclear Power Units by Nation

Country	In Operation Number of Units	In Operation Capacity Net MWe	Under Construction or on Order Number of Units	Under Construction or on Order Capacity Net MWe	Nuclear Power Production GWh*	Shutdown
Argentina	2	935	1	692	5,894	0
Armenia	1	375	0	0	2,357	1[P]
Belgium	7	5,927	0	0	45,942	1[P]
Brazil	2	1,884	1	1,245	14,795	0
Bulgaria	2	1,906	2	1,906	15,264	4[P]
Canada	18	12,604	0	0	88,318	3[P] & 4[L]
China	16	11,816	26	26,620	82,569	0
Czech Republic	6	3,678	0	0	26,696	0
Finland	4	2,716	1	1,600	22,266	0
France	58	63,130	1	1,600	423,509	12[P]
Germany	9	12,068	0	0	102,311	27[P]
Hungary	4	1,889	0	0	14,707	0
India	20	4,391	7	4,824	28,948	0
Iran	1	915	1	915	98	0
Italy	0	0	0	0	0	4[P]
Japan	50	44,215	2	2,650	156,182	9[P] & 1[L]
Kazakhstan	0	0	0	0	0	1[P]
Rep. Korea	23	20,671	3	3,640	147,763	0
Lithuania	0	0	0	0	0	2[P]
Mexico	2	1,300	0	0	9,313	0
Netherlands	1	482	0	0	3,917	1[P]
Pakistan	3	725	2	630	3,843	0
Romania	2	1,300	0	0	10,811	0
Russia	33	23,643	11	9,297	162,018	5[P]
Slovakia	4	1,816	2	782	14,342	3[P]
Slovenia	1	688	0	0	5,902	0
South Africa	2	1,830	0	0	12,939	0
Spain	8	7,567	0	0	55,121	2[P]
Sweden	10	9,331	0	0	58,098	3[P]
Switzerland	5	3,263	0	0	25,694	1[P]

Nuclear Power Units by Nation (continued)

Country	In Operation Number of Units	In Operation Capacity Net MWe	Under Construction or on Order Number of Units	Under Construction or on Order Capacity Net MWe	Nuclear Power Production GWh*	Shutdown
Ukraine	15	13,107	2	1,900	84,894	4[P]
United Kingdom	17	9,736	0	0	62,658	28[P]
United States	104	101,465	1	1,165	790,439	28[P]
Total	**436**	**370,499**	**62**	**59,245**	**2,517,980**	**139[P] & 5[L]**

* Annual electrical power production for 2011

P = Permanent Shutdown

L = Long-Term Shutdown

Note: Operable, under construction, or on order; country's short-form name used; rounded to the nearest whole number.

Sources: IAEA Power Reactor Information System Database; analysis compiled by the NRC, June 8, 2012

Nuclear Power Units by Reactor Type, Worldwide

Reactor Type	In Operation Number of Units	Net MWe
Pressurized light-water reactors (PWR)	272	250,289
Boiling light-water reactors (BWR)	84	77,726
Heavy-water reactors, all types (HWR)	47	23,140
Graphite-moderated light-water reactors (LWGR)	15	10,219
Gas-cooled reactors, all types (GCR)	16	8,545
Liquid-metal-cooled fast-breeder reactors (FBR)	2	580
Total	**436**	**370,499**

Note: MWe values rounded to the nearest whole number.

Source: IAEA Power Reactor Information System Database, www.iaea.org

Compiled by the NRC from data available as of June 8, 2012.

APPENDIX U
Native American Reservations and Trust Land within a 50-Mile Radius of a Nuclear Power Plant

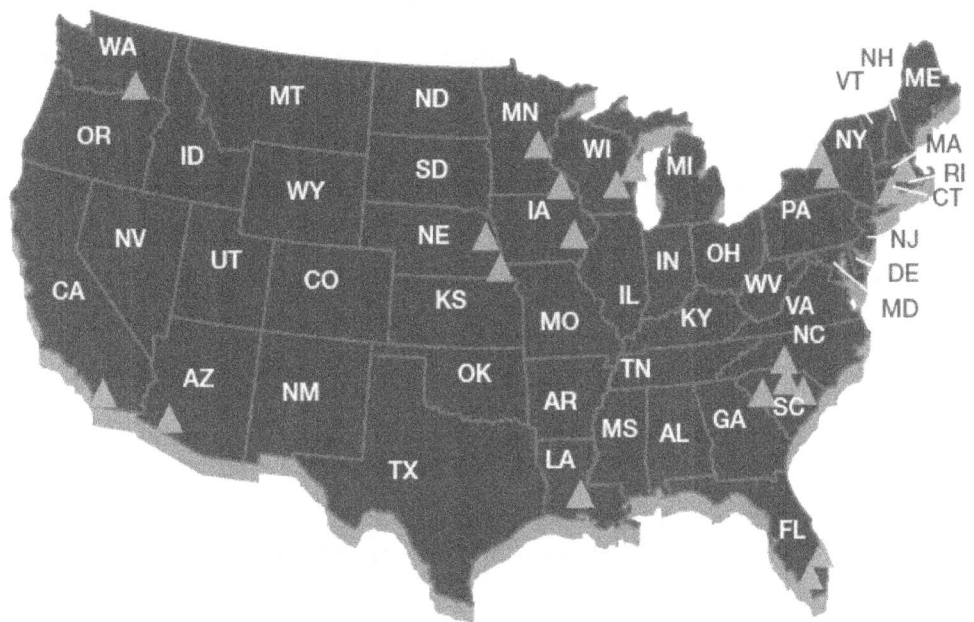

ARIZONA
Palo Verde
Ak-Chin Indian Community
Tohono O'odham
   Trust Land
Gila River Reservation
Maricopa Reserve

CALIFORNIA
San Onofre
Pechanga Reservation
   of Luiseño Indians
Pala Reservation
Pauma & Yuima Reserve
Rincon Reservation
San Pasqual Reservation
La Jolla Reservation
Cahuilla Reservation
Soboba Reservation
Santa Ysabel
Mesa Grande Reservation
Barona Reservation

CONNECTICUT
Millstone
Mohegan Reservation
Mashantucket Pequot
   Reservation
Narragansett
   Reservation

FLORIDA
St. Lucie
Brighton Reservation
   (Seminole Tribes
   of Florida)
Fort Pierce Reservation

Turkey Point
Miccosukee
   Reservation
Hollywood Reservation
   (Seminole Tribes
   of Florida)

IOWA
Duane Arnold
Sac & Fox Trust Land
Sac & Fox Reserve

LOUISIANA
River Bend
Tunica-Biloxi Reservation

MASSACHUSETTS
Pilgrim
Wampanoag
   Tribe of Grey Head
   (Aquinnah)
   Trust Land

MINNESOTA
Monticello
Shakopee Community
Shakopee Trust Land
Mille Lacs Reservation

Prairie Island
Prairie Island Community
Prairie Island Trust Land
Shakopee Community
Shakopee Trust Land

NEBRASKA
Cooper
Sac & Fox Trust Land
Sac & Fox Reservation
Kickapoo

Fort Calhoun
Winnebago Trust Land
Omaha Reservation
Winnebago Reservation

NEW YORK
FitzPatrick
Onondaga Reservation
Oneida Reservation

Nine Mile Point
Onondaga Reservation
Oneida Reservation

NORTH CAROLINA
McGuire
Catawba Reservation

SOUTH CAROLINA
Catawba
Catawba Reservation

Oconee
Eastern Cherokee
   Reservation

Summer
Catawba Reservation

WASHINGTON
Columbia
Yakama Reservation
Yakama Trust

WISCONSIN
Kewaunee
Oneida Trust Land
Oneida Reservation

Point Beach
Oneida Trust Land
Oneida Reservation

Note: This table uses NRC-abbreviated reactor names and Native American Reservation and Trust land names.

APPENDIX V
Regulatory Research Cooperative Agreements and Grants

Electric Power Research Institute	Research on central and eastern United States seismic hazards and irradiation-assisted stress-corrosion cracking
Pennsylvania State University	Assistance with cladding hydride reorientation and fracture behavior; TRACE development
International Commission on Radiological Protection	Research on radiological protection standards
Oregon State University	Research on high-temperature gas reactors
University of Maryland	Research on improved human reliability analysis methods and the cause-defense approach to common-cause failure modeling
University of California-Berkeley	Work on ground motion prediction models for central and eastern North America and postliquefaction residual strength
University of South Carolina	Research on aging electric cables and gas accumulation detection in nuclear power plants
University of Wisconsin	Research on advanced gas-cooled reactors
Texas A&M	Research on bypass flow in prismatic reactor blocks
American Nuclear Society	Support for the development and maintenance of probabilistic risk assessment (PRA)-related standards
ASME Standards Technology, LLC	Support in the following areas: Committee on Nuclear Risk Management on PRA standards, nuclear risk management, code comparison for the Multinational Design Evaluation Program, and a nondestructive examination certification program
National Academy of Sciences	Perform a study on the cancer risk for populations surrounding nuclear power plant facilities and research to develop a consensus on the assessment of soil liquefaction potential and the related infrastructure consequences
University of Tennessee	Research on seismic hazards and associated ground motion for the East Tennessee Seismic Zone
Massachusetts Institute of Technology	Research on encorporating a systems-based hazards analysis technique to support the review of digital safety systems
University of Toronto, Ontario	Research to develop a tool to confirm safety margins for modular steel-concrete composite constructions under seismic loads

APPENDIX W
Significant Enforcement Actions Issued, 2011

Issued Significant Enforcement Actions, referred to as "escalated," include notices of violation for severity level (NOVSL) I, II, or III violations; notices of violation (NOV) associated with inspection findings (NOVF) that the significance determination process (SDP) categorizes as white, yellow, or red; civil penalties (CVP); and Commission orders (CO). Escalated enforcement actions are issued to reactor, materials, and individual licensees; nonlicensees; and fuel cycle facility licensees.

Action #	Name	Type	Issue Date	Enforcement Action
EA-10-257	Carolina Power and Light Company (H. B. Robinson)	Reactor	1/31/11	NOV white SDP finding result of plant inspections
EA-10-077	Superior Well Services, Inc.	Materials	2/8/11	CO result of an alternative dispute resolution mediation - $17,000
EA-10-272	Carro & Carro Enterprises, Inc.	Nonlicensee	2/11/11	NOV SLIII
EA-11-008	Bristol Hospital, Inc.	Materials	2/17/11	NOV SLIII
EA-10-100	Mattingly Testing Services, Inc.	Materials	2/22/11	Atomic Safety and Licensing Board - Order
IA-09-035	Dr. Gary Kao	Individual	2/23/11	CO
IA-09-010	Gregory Desobry	Individual	2/23/11	CO
EA-10-153	Westinghouse Electric Company (Commercial Nuclear Fuels Division)	Fuel Cycle Facility	2/25/11	NOV SLIII
EA-11-010	Oakwood Hospital - Annapolis Center	Materials	3/4/11	NOV SLIII
EA-11-014	Exelon Generation Company, LLC (Byron)	Reactor	3/14/11	NOV white SDP finding result of plant inspections
IA-11-012	Roger A. Shaffer	Individual	3/18/11	NOV SLIII
EA-11-027	West Virginia University Hospitals, Inc.	Materials	3/25/11	NOV SLIII
EA-11-016	Community Hospitals of Indiana	Materials	4/20/11	NOV SLIII
EA-11-018	Tennessee Valley Authority (Browns Ferry)	Reactor	5/9/11	NOV red SDP finding result of plant inspections
EA-11-009	Del Valle Group	Materials	5/11/11	NOV SLIII
EA-11-037	Providence Hospital	Materials	5/17/11	NOV SLIII
EA-10-231	Alaska Industrial X-Ray, Inc.	Materials	6/7/11	CO result of an alternative dispute resolution mediation - $1,000
EA-11-094	Mercy Hospital	Materials	6/8/11	NOV SLIII
EA-11-100	Owensby and Kritikos, Inc.	Materials	6/8/11	NOV SLIII
EA-11-024	Nebraska Public Power District (Cooper)	Reactor	6/10/11	NOV white SDP finding result of plant inspections
EA-11-088	Henry Ford Macomb Hospital	Materials	6/24/11	NOV SLIII
EA-11-115	Charleston Radiation Therapy Consultants, PLLC	Materials	6/30/11	NOV SLIII
EA-11-022	Luzenac America, Inc.	Materials	7/7/11	NOV SLIII CVP-$8,500
EA-10-258	Bozeman Deaconess Hospital	Materials	7/8/11	CO result of an alternative dispute resolution mediation - $3,500
EA-11-025	Omaha Public Power District (Fort Calhoun)	Reactor	7/18/11	NOV white SDP finding result of plant inspections
EA-11-109	Liberty Hospital	Materials	7/22/11	NOV SLIII

Significant Enforcement Actions Issued, 2011 (continued)

EA-10-129	U. S. Department of the Army	Materials	8/1/11	NOV SLIII
EA-11-083	Southern California Edison Co. (San Onofre)	Reactor	8/4/11	NOV SLIII
EA-11-047	Dominion Nuclear Connecticut, Inc. (Millstone)	Reactor	8/8/11	NOV white SDP finding result of plant inspections
EA-11-056	U. S. Enrichment Corporation - Paducah Facility	Fuel Cycle Facility	8/17/11	CO result of an alternative dispute resolution mediation
EA-11-110	Northern States Power Company (Prairie Island)	Reactor	8/17/11	NOV white SDP finding result of plant inspections
EA-10-161	Professional Service Industries, Inc.	Materials	8/18/11	CO result of an alternative dispute resolution mediation - $15,000
EA-11-096	Entergy Operations, Inc. (River Bend)	Reactor	8/24/11	CO result of an alternative dispute resolution mediation
EA-11-148	First Energy Nuclear Operating Company (Perry)	Reactor	8/25/11	NOV white SDP finding result of plant inspections
EA-11-145	Carmeuse Lime, Inc.	Materials	9/2/11	NOV SLIII
EA-11-163	William Beaumont Hospital	Materials	9/2/11	NOV SLIII
EA-11-165	Crittenton Hospital	Materials	9/2/11	NOV SLIII
EA-11-179	Associated Specialists, Inc.	Materials	9/21/11	NOV SLIII
IA-11-056	Craig M. Rice	Individual	9/21/11	NOV SLIII
EA-11-061	Escanaba Paper Company	Materials	10/17/11	NOV SLIII
EA-11-209	Warner Brothers, LLC	Materials	11/8/11	NOV SLIII
EA-11-146	Cardinal Health PET Manufacturing Services, Inc.	Materials	11/9/11	NOV SLIII
IA-11-037	Christopher A. Moore	Individual	11/9/11	NOV SLIII
EA-11-095	Global Nuclear Fuel - America	Fuel Cycle Facility	11/14/11	NOV SLIII CVP-$17,500
EA-11-174	Entergy Nuclear Operations (Pilgrim)	Reactor	11/21/11	NOV white SDP finding result of plant inspections
EA-11-226	Duke Energy Carolinas (Oconee)	Reactor	12/6/11	NOV yellow SDP finding result of plant inspections
EA-11-221	Exelon Generation Company (Limerick)	Reactor	12/8/11	NOV white SDP finding result of plant inspections
EA-11-043	Accurate NDE and Inspection, LC	Materials	12/19/11	CO result of an alternative dispute resolution mediation - $13,500
EA-11-086	International Cyclotron, Inc.	Materials	12/19/11	NOV SLIII CVP-$7,000 and CO suspending license
EA-11-208	Progress Energy (Crystal River)	Reactor	12/20/11	NOV white SDP finding result of plant inspections
EA-11-251	Carolina Power and Light Company (Brunswick)	Reactor	12/27/11	NOV white SDP finding result of plant inspections

Note: Reactor facilities in a decommissioning status are listed as materials licensees. The NRC report on Issued Significant Enforcement Actions can be found on the NRC Web site at www.nrc.gov/about-nrc/regulatory/enforcement/current.html.

Quick-Reference Metric Conversion Tables

SPACE AND TIME

Quantity	From Inch-Pound Units	To Metric Units	Multiply by
Length	mi (statute)	km	1.609 347
	yd	m	*0.914 4
	ft (int)	m	*0.304 8
	in	cm	*2.54
Area	mi^2	km^2	2.589 998
	acre	m^2	4 046.873
	yd^2	m^2	0.836 127 4
	ft^2	m^2	*0.092 903 04
	in^2	cm^2	*6.451 6
Volume	acre foot	m^3	1 233.489
	yd^3	m^3	0.764 554 9
	ft^3	m^3	0.028 316 85
	ft^3	L	28.316 85
	gal	L	3.785 412
	fl oz	mL	29.573 53
	in^3	cm^3	16.387 06
Velocity	mi/h	km/h	1.609 347
	ft/s	m/s	*0.304 8
Acceleration	ft/s^2	m/s^2	*0.304 8

NUCLEAR REACTION AND IONIZING RADIATION

Quantity	From Inch-Pound Units	To Metric Units	Multiply by
Activity (of a radionuclide)	curie (Ci)	MBq	*37,000.0
	dpm	becquerel (Bq)	0.016 667
Absorbed dose	rad	gray (Gy)	*0.01
	rad	cGy	*1.0
Dose equivalent	rem	sievert (Sv)	*0.01
	rem	mSv	*10.0
	mrem	mSv	*0.01
	mrem	µSv	*10.0
Exposure (X-rays and gamma rays)	roentgen (R)	C/kg (coulomb)	0.000 258

Quick-Reference Metric Conversion Tables (continued)

HEAT

Quantity	From Inch-Pound Units	To Metric Units	Multiply by
Thermodynamic temperature	°F	K	*K = (°F + 59.67)/1.8
Celsius temperature	°F	°C	*°C = (°F–32)/1.8
Linear expansion coefficient	1/°F	1/K or 1/°C	*1.8
Thermal conductivity	(Btu • in)/(ft² • h • °F)	W/(m • °C)	0.144 227 9
Coefficient of heat transfer	Btu / (ft² • h • °F)	W/(m² • °C)	5.678 263
Heat capacity	Btu/°F	kJ/°C	1.899 108
Specific heat capacity	Btu/(lb • °F)	kJ/(kg • °C)	*4.186 8
Entropy	Btu/°F	kJ/°C	1.899 108
Specific entropy	Btu/(lb • °F)	kJ/(kg • °C)	*4.186 8
Specific internal energy	Btu/lb	kJ/kg	*2.326

MECHANICS

Quantity	From Inch-Pound Units	To Metric Units	Multiply by
Mass (weight)	ton (short)	t (metric ton)	*0.907 184 74
	lb (avdp)	kg	*0.453 592 37
Moment of mass	lb • ft	kg • m	0.138 255
Density	ton (short)/yd³	t/m³	1.186 553
	lb/ft³	g/m³	16.018 46
Concentration (mass)	lb/gal	g/L	119.826 4
Momentum	lb • ft/s	kg • m/s	0.138 255
Angular momentum	lb • ft²/s	kg • m²/s	0.042 140 11
Moment of inertia	lb • ft²	kg • m²	0.042 140 11
Force	kip (kilopound)	kN (kilonewton)	4.448 222
	lbf	N (newton)	4.448 222
Moment of force, torque	lbf • ft	N • m	1.355 818
	lbf • in	N • m	0.122 984 8
Pressure	atm (std)	kPa (kilopascal)	*101.325
	bar	kPa	*100.0
	lbf/in² (formerly psi)	kPa	6.894 757
	inHg (32 °F)	kPa	3.386 38
	ftH₂O (39.2 °F)	kPa	2.988 98
	inH₂O (60 °F)	kPa	0.248 84
	mmHg (0 °C)	kPa	0.133 322

Appendix

Quick-Reference Metric Conversion Tables (continued)

MECHANICS (continued)

Quantity	From Inch-Pound Units	To Metric Units	Multiply by
Stress	kip/in² (formerly ksi)	MPa	6.894 757
	lbf/in² (formerly psi)	MPa	0.006 894 757
	lbf/in² (formerly psi)	kPa	6.894 757
	lbf/ft²	kPa	0.047 880 26
Energy, work	kWh	MJ	*3.6
	calth	J (joule)	*4.184
	Btu	kJ	1.055 056
	ft • lbf	J	1.355 818
	therm (US)	MJ	105.480 4
Power	Btu/s	kW	1.055 056
	hp (electric)	kW	*0.746
	Btu/h	W	0.293 071 1

Note: The information contained in this table is intended to provide familiarization with commonly used SI units and provide a quick reference to aid in the understanding of documents containing SI units. The conversion factors provided have not been approved as NRC guidelines for the development of licensing actions, regulations, or policy.

To convert from metric units to inch-pound units, divide the metric unit by the conversion factor.

* Exact conversion factors

Source: Federal Standard 376B (January 27, 1993), "Preferred Metric Units for General Use by the Federal Government," and International Commission on Radiation Units and Measurements, ICRU Report 33 (1980), "Radiation Quantities and Units"

Glossary (Abbreviations and Terms Defined)

Agreement State

A State that has signed an agreement with the U.S. Nuclear Regulatory Commission (NRC) authorizing the State to regulate certain uses of radioactive materials within the State.

Atomic energy

The energy that is released through a nuclear reaction or radioactive decay process. Of particular interest is the process known as fission, which occurs in a nuclear reactor and produces energy, usually in the form of heat. In a nuclear power plant, this heat is used to boil water to produce steam that can be used to drive large turbines. This, in turn, activates generators to produce electrical power. Atomic energy is more correctly called nuclear energy.

Background radiation

The natural radiation that is always present in the environment. It includes cosmic radiation that comes from the sun and stars, terrestrial radiation that comes from the Earth, and internal radiation that exists in all living things. The typical average individual exposure in the United States from natural background sources is about 300 millirems per year.

Boiling-water reactor (BWR)

A common nuclear power reactor design in which water flows upward through the core, where it is heated by fission and allowed to boil in the reactor vessel. The resulting steam then drives turbines, which activate generators to produce electrical power. BWRs operate similarly to electrical plants using fossil fuel, except that the BWRs are powered by 370–800 nuclear fuel assemblies in the reactor core.

Brachytherapy

A nuclear medicine procedure during which a sealed radioactive source is implanted directly into a person being treated for cancer (usually of the mouth, breast, lung, prostate, ovaries, or uterus). The radioactive implant may be temporary or permanent, and the radiation attacks the tumor as long as the device remains in place. Brachytherapy uses radioisotopes, such as iridium-192 or iodine-125, which are regulated by the NRC and its Agreement States.

Byproduct material

As defined by NRC regulations, byproduct material includes any radioactive material (except enriched uranium or plutonium) produced by a nuclear reactor. It also includes the tailings or wastes produced by the extraction or concentration of uranium or thorium or the fabrication of fuel for nuclear reactors. Additionally, it is any material that has been made radioactive through the use of a particle accelerator or any discrete source of radium-226 used for a commercial, medical, or research activity. In addition, the NRC, in consultation with the U.S. Environmental Protection Agency (EPA), U.S. Department of Energy (DOE), U.S. Department of Homeland Security (DHS), and others, can designate as byproduct material any source of naturally occurring radioactive material, other than source material, that it determines would pose a threat to public health and safety or the common defense and security of the United States.

Canister

See *Dry cask storage*.

Capability

The maximum load that a generating unit, generating station, or other electrical apparatus can carry under specified conditions for a given period of time without exceeding approved limits of temperature and stress.

Capacity

The amount of electric power that a generating unit can produce. The amount of electric power that a generator, turbine transformer, transmission, circuit, or system is able to produce, as rated by the manufacturer.

Capacity charge

One of two elements in a two-part pricing method used in capacity transactions (the other element is the energy charge). The capacity charge, sometimes called the demand charge, is assessed on the capacity (amount of electric power) being purchased.

Capacity factor

The ratio of the available capacity (the amount of electrical power actually produced by a generating unit) to the theoretical capacity (the amount of electrical power that could theoretically have been produced if the generating unit had operated continuously at full power) during a given time period.

Capacity utilization

A percentage representing the extent to which a generating unit fulfilled its capacity in generating electric power over a given time period. This percentage is defined as the margin between the unit's available capacity (the amount of electrical power the unit actually produced) and its theoretical capacity (the amount of electrical power that could have been produced if the unit had operated continuously at full power) during a certain time period. Capacity utilization is computed by dividing the amount actually produced by the theoretical capacity and multiplying by 100.

Cask

A heavily shielded container used for the dry storage or shipment (or both) of radioactive materials such as spent nuclear fuel or other high-level radioactive waste (HLW). Casks are often made from lead, concrete, or steel. Casks must meet regulatory requirements and are not intended for long-term disposal in a repository.

Classified information

Information that could be used by an adversary to harm the United States or its allies and thus must be protected. The NRC has two types of classified information. The first type, known as national security information, is information that is classified by an Executive order. Its release would damage national security to some degree. The second type, known as restricted data, is information that is classified by the Atomic Energy Act of 1954, as amended. It would assist individuals or organizations in designing, manufacturing, or using nuclear weapons. Access to both types of information is restricted to authorized persons who have been properly cleared and have a "need to know" the information for their official duties.

Combined license (COL)

An NRC-issued license that authorizes a licensee to construct and (with certain specified conditions) operate a nuclear power plant at a specific site, in accordance with established laws and regulations. A COL is valid for 40 years (with the possibility of a 20-year renewal).

Commercial sector (energy users)

Generally, nonmanufacturing business establishments, including hotels, motels, and restaurants; wholesalers and retail stores; and health, social, and educational institutions. However, utilities may categorize commercial service as all consumers whose demand or annual usage exceeds some specified limit that is categorized as residential service.

Compact

A group of two or more States that have formed business alliances to dispose of low-level radioactive waste (LLW) on a regional basis.

Construction recapture

The maximum number of years that could be added to a facility's license expiration date to recapture the period between the date the NRC issued the facility's construction permit and the date it granted an operating license. A licensee must submit an application to request this extension.

Containment structure

A gas-tight shell or other enclosure around a nuclear reactor to confine fission products that otherwise might be released to the atmosphere in the event of an accident. Such enclosures are usually dome-shaped and made of steel-reinforced concrete.

Contamination

Undesirable radiological, chemical, or biological material (with a potentially harmful effect) that is either airborne or deposited in (or on the surface of) structures, objects, soil, water, or living organisms in a concentration that makes the medium unfit for its next intended use.

Criticality

The normal operating condition of a reactor, in which nuclear fuel sustains a fission chain reaction. A reactor achieves criticality (and is said to be critical) when each fission event releases a sufficient number of neutrons to sustain an ongoing series of reactions.

Decommissioning

The process of safely closing a nuclear power plant (or other facility where nuclear materials are handled) to retire it from service after its useful life has ended. This process primarily involves decontaminating the facility to reduce residual radioactivity and then releasing the property for unrestricted or (under certain conditions) restricted use. This often includes dismantling the facility or dedicating it to other purposes. Decommissioning begins after the nuclear fuel, coolant, and radioactive waste are removed.

DECON

A method of decommissioning, in which structures, systems, and components that contain radioactive contamination are removed from a site and safely disposed of at a commercially operated LLW disposal facility or decontaminated to a level that permits the site to be released for unrestricted use shortly after it ceases operation.

Decontamination

A process used to reduce, remove, or neutralize radiological, chemical, or biological contamination to reduce the risk of exposure. Decontamination may be accomplished by cleaning or treating surfaces to reduce or remove the contamination; filtering contaminated air or water; subjecting contamination to evaporation and precipitation; or covering the contamination to shield or absorb the radiation. The process can also simply allow adequate time for natural radioactive decay to decrease the radioactivity.

Defense in depth

An approach to designing and operating nuclear facilities that prevents and mitigates accidents that release radiation or hazardous materials. The key is creating multiple independent and redundant layers of defense to compensate for potential human and mechanical failures so that no single layer, no matter how robust, is exclusively relied upon. Defense in depth includes the use of access controls, physical barriers, redundant and diverse key safety functions, and emergency response measures.

Depleted uranium

Uranium with a percentage of uranium-235 lower than the 0.7 percent (by mass) contained in natural uranium. (The normal residual uranium-235 content in depleted uranium is 0.2–0.3 percent, with uranium-238 comprising the remaining 98.7–98.8 percent.) Depleted uranium is the byproduct of the uranium enrichment process. Depleted uranium can be blended with highly enriched uranium, such as that from weapons, to make reactor fuel.

Design-basis threat (DBT)

A profile of the type, composition, and capabilities of an adversary. The NRC uses the DBT as a basis for designing safeguards systems to protect against acts of radiological sabotage and to prevent the theft of special nuclear material. Nuclear facility licensees are expected to demonstrate they can defend against the DBT.

Design certification

Certification and approval by the NRC of a standard nuclear power plant design independent of a specific site or an application to construct or operate a plant. A design certification is valid for 15 years from the date of issuance but can be renewed for an additional 10 to 15 years.

Dry cask storage

A method for storing spent nuclear fuel above ground in special containers known as casks. After fuel has been cooled in a spent fuel pool for at least 1 year, dry cask storage allows approximately one to six dozen spent fuel assemblies to be sealed in casks and surrounded by inert gas. The casks are large, rugged cylinders made of steel or steel-reinforced concrete (18 or more inches thick or 45.72 or more centimeters). They are welded or bolted closed, and each cask is surrounded by steel, concrete, lead, or other material to provide leak-tight containment and radiation shielding. The casks may be placed horizontally in aboveground concrete bunkers or vertically in concrete vaults or on concrete pads.

Early site permit (ESP)

A permit through which the NRC resolves site safety, environmental protection, and emergency preparedness (EP) issues to approve one or more proposed sites for a nuclear power facility, independent of a specific nuclear plant design or an application for a construction permit or COL. An ESP is valid for 10 to 20 years but can be renewed for an additional 10 to 20 years.

Economic Simplified Boiling-Water Reactor (ESBWR)

A 4,500-megawatts thermal nuclear reactor design, which has passive safety features and uses natural circulation (with no recirculation pumps or associated piping) for normal operation. GE-Hitachi Nuclear Energy (GEH) submitted an application for final design approval and standard design certification for the ESBWR on August 24, 2005.

Efficiency, plant

The percentage of the total energy content of a power plant's fuel that is converted into electricity. The remaining energy is lost to the environment as heat.

Electric power grid

A system of synchronized power providers and consumers, connected by transmission and distribution lines and operated by one or more control centers. In the continental United States, the electric power grid consists of three systems—the Eastern Interconnect, the Western Interconnect, and the Texas Interconnect. In Alaska and Hawaii, several systems encompass areas smaller than the State.

Electric utility

A corporation, agency, authority, person, or other legal entity that owns or operates facilities within the United States, its territories, or Puerto Rico for the generation, transmission, distribution, or sale of electric power (primarily for use by the public). Facilities that qualify as cogenerators or small power producers under the Public Utility Regulatory Policies Act are not considered electric utilities.

Emergency classifications

Sets of plant conditions that indicate various levels of risk to the public and that might require response by an offsite emergency response organization to protect citizens near the site.

Emergency preparedness

The programs, plans, training, exercises, and resources necessary to prepare emergency personnel to rapidly identify, evaluate, and react to emergencies, including those arising from terrorism or natural events such as hurricanes. EP strives to ensure that operators of nuclear power plant and certain fuel cycle facilities can implement measures to protect public health and safety in the event of a radiological emergency. Plant operators, as a condition of their licenses, must develop and maintain EP plans that meet NRC requirements.

Energy Information Administration (EIA)

The agency, within the U.S. Department of Energy, that provides policy-neutral statistical data, forecasts, and analyses to promote sound policymaking, efficient markets, and public understanding regarding energy and its interaction with the economy and the environment.

Enrichment

See "Uranium enrichment".

ENTOMB

A method of decommissioning, in which radioactive contaminants are encased in a structurally long-lived material, such as concrete. The entombed structure is maintained and surveillance is continued until the radioactive waste decays to a level permitting termination of the license and unrestricted release of the property. During the entombment period, the licensee maintains the license previously issued by the NRC.

Event Notification System

An automated event tracking system used internally by the NRC's Headquarters Operations Center to track incoming notifications of significant nuclear events with an actual or potential effect on the health and safety of the public and the environment. Significant events are reported to the Operations Center by the NRC's licensees, Agreement States, other Federal agencies, the public, and other stakeholders.

Exposure

Absorption of ionizing radiation or ingestion of a radioisotope. Acute exposure is a large exposure received over a short period of time. Chronic exposure is exposure received over a long period of time, such as during a lifetime. The National Council on Radiation Protection and Measurements estimates that an average person in the United States receives a total annual dose of about 0.62 rem (620 millirem) from all radiation source, a level that has not been shown to cause humans any harm. Of this total, natural background sources of radiation—including radon and thoron gas, natural radiation from soil and rocks, radiation from space, and radiation sources that are found naturally within the human body—account for approximately 50 percent. Medical procedures such as computed tomography (CT scans) and nuclear medicine account for approximately another 48 percent. Other small contributors of exposure to the U.S. population include consumer products and activities, industrial and research uses, and occupational tasks. The maximum permissible yearly dose for a person working with or around nuclear material is 5 rem.

Federal Emergency Management Agency (FEMA)

A component of DHS responsible for protecting the nation and reducing the loss of life and property from all hazards, such as natural disasters and acts of terrorism. FEMA leads and supports a risk-based, comprehensive emergency management system of preparedness, protection, response, recovery, and mitigation. FEMA also administers the National Flood Insurance Program.

Federal Energy Regulatory Commission (FERC)

An independent agency that regulates the interstate transmission of electricity, natural gas, and oil. FERC also regulates and oversees hydropower projects and the construction of liquefied natural gas terminals and interstate natural gas pipelines. FERC protects the economic, environmental, and safety interests of the American public, while working to ensure abundant, reliable energy in a fair, competitive market.

Fiscal year (FY)

The 12-month period from October 1 through September 30 used by the Federal Government for budget formulation and execution. The FY is designated by the calendar year in which it ends; for example, FY 2009 runs from October 1, 2008, through September 30, 2009.

Fissile material

A nuclide that is capable of undergoing fission after capturing low-energy thermal (slow) neutrons. Although sometimes used as a synonym for fissionable material, this term has acquired its more restrictive interpretation with the limitation that the nuclide must be fissionable by thermal neutrons. With that interpretation, the three primary fissile materials are uranium-233, uranium-235, and plutonium-239. This definition excludes natural uranium and depleted uranium that have not been irradiated or have only been irradiated in thermal reactors.

Fission (fissioning)

The splitting of an atom, which releases a considerable amount of energy (usually in the form of heat) that can be used to produce electricity. Fission may be spontaneous but is usually caused by the nucleus of an atom becoming unstable (or "heavy") after capturing or absorbing a neutron. During fission, the heavy nucleus splits into roughly equal parts, producing the nuclei of at least two lighter elements. In addition to energy, this reaction usually releases gamma radiation and two or more daughter neutrons.

Force on force

Inspections designed to evaluate and improve the effectiveness of a licensee's security force and ability to defend a nuclear power plant and other nuclear facilities against a DBT. An essential part of the security program instituted by the NRC, a full force-on-force inspection spans 2 weeks and includes tabletop drills and multiple simulated combat exercises between a mock commando-type adversary and the plant's security force.

Foreign Assignee Program

An on-the-job training program, sponsored by the NRC for assignees from other countries, usually under bilateral information exchange arrangements with their respective regulatory organizations.

Freedom of Information Act (FOIA)

A Federal law that requires Federal agencies to provide, upon written request, access to records or information. Some material is exempt from FOIA, and FOIA does not apply to records that are maintained by State and local governments, Federal contractors, grantees, or private organizations or businesses.

Fuel assembly (fuel bundle, fuel element)

A structured group of fuel rods (long, slender, metal tubes containing pellets of fissionable material, which provide fuel for nuclear reactors). Depending on the design, each reactor vessel may have dozens of fuel assemblies (also known as fuel bundles), each of which may contain 200 or more fuel rods.

Fuel cycle

The series of steps involved in supplying fuel for nuclear power reactors includes the following:

- uranium recovery to extract (or mine) uranium ore and concentrate (or mill) the ore to produce yellowcake

- conversion of yellowcake into uranium hexafluoride (UF_6)

- enrichment to increase the concentration of uranium-235 in UF_6

- fuel fabrication to convert enriched UF_6 into fuel for nuclear reactors

- use of the fuel in reactors (nuclear power, research, or naval propulsion)

- interim storage of spent nuclear fuel

- reprocessing of HLW to recover the fissionable material remaining in the spent fuel (currently not done in the United States)

- final disposition (disposal) of HLW

The NRC regulates these processes, as well as the fabrication of mixed oxide (MOX) nuclear fuel, which is a combination of uranium and plutonium oxides.

Fuel reprocessing (recycling)

The processing of reactor fuel to separate the unused fissionable material from waste material. Reprocessing extracts isotopes from spent nuclear fuel so they can be used again as reactor fuel. Commercial reprocessing is not practiced in the United States, although it has been practiced in the past. However, the U.S. Department of Defense oversees reprocessing programs at DOE facilities such as in Hanford, WA, and Savannah River, SC. These wastes, as well as those wastes at a formerly operating commercial reprocessing facility at West Valley, NY, are not regulated by the NRC.

Fuel rod

A long, slender, zirconium metal tube containing pellets of fissionable material, which provide fuel for nuclear reactors. Fuel rods are assembled into bundles called fuel assemblies, which are loaded individually into the reactor core.

Full-time equivalent (FTE)

A human resources measurement equal to one staff person working full time for 1 year.

Gas centrifuge

A uranium enrichment process used to prepare uranium for use in fabricating fuel for nuclear reactors by separating its isotopes (as gases) based on their slight difference in mass. This process uses a large number of interconnected centrifuge machines (rapidly spinning cylinders). URENCO operates a gas centrifuge enrichment facility in New Mexico, and USEC and AREVA have received licenses to construct and operate facilities in Ohio and Idaho, respectively.

Gas chromatography

A way of separating chemical substances from a mixed sample by passing the sample, carried by a moving stream of gas, through a tube packed with a finely divided solid that may be coated with a liquid film. Gas chromatography devices are used to analyze air pollutants, blood alcohol content, essential oils, and food products.

Gaseous diffusion

A uranium enrichment process used to prepare uranium for use in fabricating fuel for nuclear reactors by separating its isotopes (as gases) based on their slight difference in velocity. (Lighter isotopes diffuse faster through a porous membrane or vessel than do heavier isotopes.) This process involves filtering UF_6 gas to separate uranium-234 and uranium-235 from uranium-238, increasing the percentage of uranium-235 from 1 to 3 percent. The only gaseous diffusion plant in operation in the United States is in Paducah, KY, and it enriches to 5 percent. A similar plant near Piketon, OH, was closed in March 2001. Both plants are leased by USEC from DOE and have been regulated by the NRC since March 4, 1997.

Gauging devices

Devices used to measure, monitor, and control the thickness of sheet metal, textiles, paper napkins, newspaper, plastics, photographic film, and other products as they are manufactured. Gauges mounted in fixed locations are designed for measuring or controlling material density, flow, level, thickness, or weight. The gauges contain sealed sources that radiate through the substance being measured to a readout or controlling device. Portable gauging devices, such as moisture density gauges, are used at field locations. These gauges contain a gamma-emitting sealed source, usually cesium-137, or a sealed neutron source, usually americium-241 or beryllium.

Generation (gross)

The total amount of electric energy produced by a generating station, as measured at the generator terminals.

Generation (net)

The gross amount of electric energy produced by a generating station, minus the amount used to operate the station. Net generation is usually measured in watthours.

Generator capacity

The maximum amount of electric energy that a generator can produce (from the mechanical energy of the turbine), adjusted for ambient conditions. Generator capacity is commonly expressed in megawatts (MW).

Generator nameplate capacity

The maximum amount of electric energy that a generator can produce under specific conditions, as rated by the manufacturer. Generator nameplate capacity is usually expressed in kilovolt-amperes and kilowatts (kW), as indicated on a nameplate that is physically attached to the generator.

Geological repository

An excavated, underground facility that is designed, constructed, and operated for safe and secure permanent disposal of HLW. A geological repository uses an engineered barrier system and a portion of the site's natural geology, hydrology, and geochemical systems to isolate the radioactivity of the waste. The Nuclear Waste Policy Act of 1982, as amended, specified that this waste would be disposed of in a deep geologic repository, and that Yucca Mountain, NV, would be the single candidate site for such a repository. On June 3, 2008, DOE submitted a license application to the NRC seeking authorization to construct the Yucca Mountain repository. On January 29, 2010, the President created the Blue Ribbon Commission on America's Nuclear Future to reassess the national policy on HLW disposal.

Gigawatt (GW)

A unit of power equivalent to one billion (1,000,000,000) watts.

Gigawatthour (GWh)

One billion (1,000,000,000) watthours.

Grid

See *Electric power grid*.

Half-life (radiological)

The time required for half the atoms of a particular radioisotope to decay into another isotope that has half the activity of the original radioisotope. A specific half-life is a characteristic property of each radioisotope. Measured half-lives range from millionths of a second to billions of years, depending on the stability of the nucleus. Radiological half-life is related to, but different from, the biological half-life and the effective half-life.

Health physics

The science concerned with recognizing and evaluating the effects of ionizing radiation on the health and safety of people and the environment, monitoring radiation exposure, and controlling the associated health risks and environmental hazards to permit the safe use of technologies that produce ionizing radiation.

Glossary

High-level radioactive waste (HLW)

The highly radioactive materials produced as byproducts of fuel reprocessing or of the reactions that occur inside nuclear reactors. HLW includes the following:

- irradiated spent nuclear fuel discharged from commercial nuclear power reactors

- the highly radioactive liquid and solid materials resulting from the reprocessing of spent nuclear fuel, which contain fission products in concentration (this includes some reprocessed HLW from defense activities and a small quantity of reprocessed commercial HLW)

- other highly radioactive materials that the Commission may determine require permanent isolation

Highly (or High-) enriched uranium

Uranium enriched to at least 20 percent uranium-235 (a higher concentration than exists in natural uranium ore).

In situ recovery (ISR)

One of the two primary recovery methods that are currently used to extract uranium from ore bodies where they are normally found underground (in other words, in situ), without physical excavation. ISR is also known as "solution mining" or in situ leaching.

Incident response

Activities that address the short-term, direct effects of a natural or human-caused event and require an emergency response to protect life or property.

Independent spent fuel storage installation (ISFSI)

A complex designed and constructed for the interim storage of spent nuclear fuel; solid, reactor-related, greater than Class C waste; and other associated radioactive materials. A spent fuel storage facility may be considered independent, even if it is located on the site of another NRC-licensed facility.

International Atomic Energy Agency (IAEA)

The IAEA is the world's center of cooperation in the nuclear field. It was set up in 1957 as the world's "Atoms for Peace" organization within the United Nations family. The agency works with its 154 member States and multiple partners worldwide to promote safe, secure, and peaceful nuclear technology.

International Nuclear Regulators Association

An association established in January 1997 to give international nuclear regulators a forum to discuss nuclear safety. Countries represented include Canada, France, Japan, the Republic of South Korea, Spain, Sweden, the United Kingdom, and the United States.

Irradiation

Exposure to ionizing radiation. Irradiation may be intentional, such as in cancer treatments or in sterilizing medical instruments. Irradiation may also be accidental, such as from exposure to an unshielded source. Irradiation does not usually result in radioactive contamination, but damage can occur, depending on the dose received.

Isotope

Two or more forms (or atomic configurations) of a given element that have identical atomic numbers (the same number of protons in their nuclei) and the same or very similar chemical properties but different atomic masses (different numbers of neutrons in their nuclei) and distinct physical properties. Thus, carbon-12, carbon-13, and carbon-14 are isotopes of the element carbon, and the numbers denote the approximate atomic masses. Among their distinct physical properties, some isotopes (known as radioisotopes) are radioactive, because their nuclei emit radiation as they strive toward a more stable nuclear configuration. For example, carbon-12 and carbon-13 are stable, but carbon-14 is unstable and radioactive.

Kilowatt (kW)

A unit of power equivalent to one thousand (1,000) watts.

Licensed material

Source material, byproduct material, or special nuclear material that is received, possessed, used, transferred, or disposed of under a general or specific license issued by the NRC or Agreement States.

Licensee

A company, organization, institution, or other entity to which the NRC has granted a general or specific license to construct or operate a nuclear facility, or to receive, possess, use, transfer, or dispose of source, byproduct, or special nuclear material.

Licensing basis

The collection of documents or technical criteria that provides the basis upon which the NRC issues a license to construct or operate a nuclear facility; to conduct operations involving the emission of radiation; or to receive, possess, use, transfer, or dispose of source, byproduct, or special nuclear material.

Light-water reactor

A term used to describe reactors using ordinary water as a coolant, including BWRs and pressurized-water reactors (PWRs), the most common types used in the United States.

Low-level radioactive waste (LLW)

A general term for a wide range of items that have become contaminated with radioactive material or have become radioactive through exposure to neutron radiation. A variety of industries, hospitals and medical institutions, educational and research institutions, private or government laboratories, and nuclear fuel cycle facilities generate LLW as part of their day-to-day use of radioactive materials. Some examples include radioactively contaminated protective shoe covers and clothing; cleaning rags, mops, filters, and reactor water treatment residues; equipment and tools; medical tubes, swabs, and hypodermic syringes; and carcasses and tissues from laboratory animals. The radioactivity in these wastes can range from just above natural background levels to much higher levels, such as seen in parts from inside the reactor vessel in a nuclear power plant. LLW is typically stored on site by licensees, either until it has decayed away and can be disposed of as ordinary trash, or until the accumulated amount becomes large enough to warrant shipment to an LLW disposal site.

Maximum dependable capacity (gross)

The maximum amount of electricity that the main generating unit of a nuclear power reactor can reliably produce during the summer or winter (usually summer, but whichever represents the most restrictive seasonal conditions, with the least electrical output). The dependable capacity varies during the year, because temperature variations in cooling water affect the unit's efficiency. Thus, this is the gross electrical output as measured (in watts unless otherwise noted) at the output terminals of the turbine generator.

Maximum dependable capacity (net)

The gross maximum dependable capacity of the main generating unit in a nuclear power reactor, minus the amount used to operate the station. Net maximum dependable capacity is measured in watts unless otherwise noted.

Megawatt (MW)

A unit of power equivalent to one million (1,000,000) watts.

Metric ton

Approximately 2,200 pounds.

Mill tailings

Primarily, the sandy process waste material from a conventional uranium recovery facility. This naturally radioactive ore residue contains the radioactive decay products from the uranium chains (mainly the uranium-238 chain) and heavy metals. Although the milling process recovers about 93 percent of the uranium, the residues (known as "tailings") contain several naturally occurring radioactive elements, including uranium, thorium, radium, polonium, and radon.

Mixed oxide (MOX) fuel

A type of nuclear reactor fuel (often called "MOX") that contains plutonium oxide mixed with either natural or depleted uranium oxide, in ceramic pellet form. (This differs from conventional nuclear fuel, which is made of pure uranium oxide.) Using plutonium reduces the amount of highly enriched uranium needed to produce a controlled reaction in commercial light-water reactors. However, plutonium exists only in trace amounts in nature and, therefore, must be produced by neutron irradiation of uranium-238 or obtained from other manufactured sources. As directed by Congress, the NRC regulates the fabrication of MOX fuel by DOE, a program that is intended to dispose of plutonium from international nuclear disarmament agreements.

Monitoring of radiation

Periodic or continuous determination of the amount of ionizing radiation or radioactive contamination in a region. Radiation monitoring is a safety measure to protect the health and safety of the public and the environment through the use of bioassay, alpha scans, and other radiological survey methods to monitor air, surface water and ground water, soil and sediment, equipment surfaces, and personnel.

National Response Framework

The guiding principles, roles, and structures that enable all domestic incident response partners to prepare for and provide a unified national response to disasters and emergencies. It describes how the Federal Government, States, Tribes, communities, and the private sector work together to coordinate a national response. The framework, which became effective March 22, 2008, builds upon the National Incident Management System, which provides a template for managing incidents.

National Source Tracking System (NSTS)

A secure, Web-based data system that helps the NRC and its Agreement States track and regulate the medical, industrial, and academic uses of certain nuclear materials, from the time they are manufactured or imported to the time of their disposal or exportation. This information enhances the ability of the NRC and Agreement States to conduct inspections and investigations, communicate information to other government agencies, and verify the ownership and use of nationally tracked sources.

Natural uranium

Uranium containing the relative concentrations of isotopes found in nature (0.7 percent uranium-235, 99.3 percent uranium-238, and a trace amount of uranium-234 by mass). In terms of radioactivity, however, natural uranium contains approximately 2.2 percent uranium-235, 48.6 percent uranium-238, and 49.2 percent uranium-234. Natural uranium can be used as fuel in nuclear reactors.

Net electric generation

The gross amount of electric energy produced by a generating station, minus the amount used to operate the station. Note: Electricity required for pumping at pumped-storage plants is regarded as electricity for station operation and is deducted from gross generation. Net electric generation is measured in watthours, except as otherwise noted.

Net summer capacity

The steady hourly output that generating equipment is expected to supply to system load, exclusive of auxiliary power, as demonstrated by measurements at the time of peak demand (summer). Net summer capacity is measured in watts unless otherwise noted.

Nonpower reactor (research and test reactor)

A nuclear reactor that is used for research, training, or development purposes (which may include producing radioisotopes for medical and industrial uses) but has no role in producing electrical power. These reactors, which are also known as research and test reactors, contribute to almost every field of science, including physics, chemistry, biology, medicine, geology, archeology, and ecology.

NRC Headquarters Operations Center

The primary center of communication and coordination among the NRC, its licensees, State and Tribal agencies, and other Federal agencies regarding operating events involving nuclear reactors or materials. Located in Rockville, MD, the Headquarters Operations Center is staffed 24 hours a day by employees trained to receive and evaluate event reports and coordinate incident response activities.

Nuclear energy

See *Atomic energy.*

Nuclear Energy Agency (NEA)

A specialized agency within the Organisation for Economic Co-operation and Development (OECD), which was created to assist its member countries in maintaining and further developing the scientific, technological, and legal bases for safe, environmentally friendly, and economical use of nuclear energy for peaceful purposes. The NEA's current membership consists of 30 countries in Europe, North America, and the Asia-Pacific region, which account for approximately 85 percent of the world's installed nuclear capacity.

Nuclear fuel

Fissionable material that has been enriched to a composition that will support a self-sustaining fission chain reaction when used to fuel a nuclear reactor, thereby producing energy (usually in the form of heat or useful radiation) for use in other processes.

Nuclear materials

See *Special nuclear material, Source material,* and *Byproduct material.*

Nuclear Material Management and Safeguards System (NMMSS)

A centralized U.S. Government database used to track and account for source and special nuclear material and used to ensure that it has not been stolen or diverted to unauthorized users. The system contains current and historical data on the possession, use, and shipment of source and special nuclear material within the United States, as well as all exports and imports of such material. The database is jointly funded by the NRC and DOE and is operated under a DOE contract.

Nuclear poison (or neutron poison)

In reactor physics, a substance (other than fissionable material) that has a large capacity for absorbing neutrons in the vicinity of the reactor core. This effect may be undesirable in some reactor applications because it may prevent or disrupt the fission chain reaction, thereby affecting normal operation. However, neutron-absorbing materials (commonly known as "poisons") are intentionally inserted into some types of reactors to decrease the reactivity of their initial fresh fuel load. (Adding poisons, such as control rods or boron, is described as adding "negative reactivity" to the reactor.)

Nuclear power plant

A thermal power plant, in which the energy (heat) released by the fissioning of nuclear fuel is used to boil water to produce steam. The steam spins the propeller-like blades of a turbine that turns the shaft of a generator to produce electricity. Of the various nuclear power plant designs, PWRs and BWRs are in commercial operation in the United States. These facilities generate about 20 percent of U.S. electrical power.

Nuclear and Radiological Incident Annex

An annex to the National Response Framework, which provides for a timely, coordinated response by Federal agencies to nuclear or radiological accidents or incidents within the United States. This annex covers radiological dispersal devices and improvised nuclear devices, as well as accidents involving commercial reactors or weapons production facilities, lost radioactive sources, transportation accidents involving radioactive material, and foreign accidents involving nuclear or radioactive material.

Nuclear reactor

The heart of a nuclear power plant or nonpower reactor, in which nuclear fission may be initiated and controlled in a self-sustaining chain reaction to generate energy or produce useful radiation. Although there are many types of nuclear reactors, they all incorporate certain essential features, including the use of fissionable material as fuel, a moderator (such as water) to increase the likelihood of fission (unless reactor operation relies on fast neutrons), a reflector to conserve escaping neutrons, coolant provisions for heat removal, instruments for monitoring and controlling reactor operation, and protective devices (such as control rods and shielding).

Nuclear waste

A subset of radioactive waste that includes unusable byproducts produced during the various stages of the nuclear fuel cycle, including extraction, conversion, and enrichment of uranium; fuel fabrication; and use of the fuel in nuclear reactors. Specifically, these stages produce a variety of nuclear waste materials, including uranium mill tailings, depleted uranium, and spent (depleted) fuel, all of which are regulated by the NRC. (By contrast, "radioactive waste" is a broader term, which includes all wastes that contain radioactivity, regardless of how they are produced. It is not considered "nuclear waste," because it is not produced through the nuclear fuel cycle and is generally not regulated by the NRC.)

Occupational dose

The internal and external dose of ionizing radiation received by workers in the course of employment in such areas as fuel cycle facilities, industrial radiography, nuclear medicine, and nuclear power plants. These workers are exposed to varying amounts of radiation, depending on their jobs and the sources with which they work. The NRC requires its licensees to limit occupational exposure to 5,000 mrem (50 millisievert) per year. Occupational dose does not include the dose received from natural background sources, doses received as a medical patient or participant in medical research programs, or "second-hand doses" to members of the public received through exposure to patients treated with radioactive materials.

Organisation for Economic Co-operation and Development (OECD)

An intergovernmental organization (based in Paris, France) that provides a forum for discussion and cooperation among the governments of industrialized countries committed to democracy and the market economy. The primary goal of OECD and its member countries is to support sustainable economic growth, boost employment, raise living standards, maintain financial stability, assist other countries' economic development, and contribute to growth in world trade. In addition, OECD is a reliable source of comparable statistics and economic and social data. OECD also monitors trends, analyzes and forecasts economic developments, and researches social changes and evolving patterns in trade, environment, agriculture, technology, taxation, and other areas.

Orphan sources (unwanted radioactive material)

Sealed sources of radioactive material contained in a small volume (but not radioactively contaminated soils and bulk metals) in any one or more of the following conditions:

- an uncontrolled condition that requires removal to protect public health and safety from a radiological threat

- a controlled or uncontrolled condition, for which a responsible party cannot be readily identified

- a controlled condition, compromised by an inability to ensure the continued safety of the material (e.g., the licensee may have few or no options to provide for safe disposition of the material)

- an uncontrolled condition, in which the material is in the possession of a person who did not seek, and is not licensed, to possess it

- an uncontrolled condition, in which the material is in the possession of a State radiological protection program solely to mitigate a radiological threat resulting from one of the above conditions, and for which the State does not have the necessary means to provide for the appropriate disposition of the material

Outage

The period during which a generating unit, transmission line, or other facility is out of service. Outages may be forced or scheduled and full or partial.

Outage (forced)

The shutdown of a generating unit, transmission line, or other facility for emergency reasons, or a condition in which the equipment is unavailable as a result of an unanticipated breakdown. An outage (whether full, partial, or attributable to a failed start) is considered "forced" if it could not reasonably be delayed beyond 48 hours from identification of the problem, if there had been a strong commercial desire to do so. In particular, the following problems may result in forced outages:

- any failure of mechanical, fuel handling, or electrical equipment or controls within the generator's ownership or direct responsibility (i.e., from the point the generator is responsible for the fuel through to the electrical connection point)

- a failure of a mine or fuel transport system dedicated to that power station with a resulting fuel shortage that cannot be economically managed

- inadvertent or operator error

- limitations caused by fuel quality

Forced outages do not include scheduled outages for inspection, maintenance, or refueling.

Outage (full forced)

A forced outage that causes a generating unit to be removed from the committed state (when the unit is electrically connected and generating or pumping) or the available state (when the unit is available for dispatch as a generator or pump but is not electrically connected and not generating or pumping). Full-forced outages do not include failed starts.

Outage (scheduled)

The shutdown of a generating unit, transmission line, or other facility for inspection, maintenance, or refueling, which is scheduled well in advance (even if the schedule changes). Scheduled outages do not include forced outages and could be deferred if there were a strong commercial reason to do so.

Pellet, fuel

A thimble-sized ceramic cylinder (approximately 3/8-inch in diameter and 5/8-inch in length), consisting of uranium (typically uranium oxide), which has been enriched to increase the concentration of uranium-235 (U-235) to fuel a nuclear reactor. Modern reactor cores in PWRs and BWRs may contain up to 10 million pellets, stacked in the fuel rods that form fuel assemblies.

Performance-based regulation

A regulatory approach that focuses on desired, measurable outcomes, rather than prescriptive processes, techniques, or procedures. Performance-based regulation leads to defined results without specific direction regarding how those results are to be obtained. At the NRC, performance-based regulatory actions focus on identifying performance measures that ensure an adequate safety margin and offer incentives for licensees to improve safety without formal regulatory intervention by the agency.

Performance indicator

A quantitative measure of a particular attribute of licensee performance that shows how well a plant is performing when measured against established thresholds. Licensees submit their data quarterly; the NRC regularly conducts inspections to verify the submittals and then uses its own inspection data plus the licensees' submittals to assess each plant's performance.

Possession-only license

A license, issued by the NRC, that authorizes the licensee to possess specific nuclear material but does not authorize its use or the operation of a nuclear facility.

Power uprate

The process of increasing the maximum power level at which a commercial nuclear power plant may operate. This power level, regulated by the NRC, is included in the plant's operating license and technical specifications. A licensee may only change its maximum power output after the NRC approves an uprate application. The NRC analyses must demonstrate that the plant could continue to operate safely with its proposed new configuration. When all requisite conditions are fulfilled, the NRC may grant the power uprate by amending the plant's operating license and technical specifications.

Pressurized-water reactor (PWR)

A common nuclear power reactor design in which very pure water is heated to a very high temperature by fission, kept under high pressure (to prevent it from boiling), and converted to steam by a steam generator (rather than by boiling, as in a BWR). The resulting steam is used to drive turbines, which activate generators to produce electrical power. A PWR essentially operates like a pressure cooker, where a lid is tightly placed over a pot of heated water, causing the pressure inside to increase as the temperature increases (because the steam cannot escape) but keeping the water from boiling at the usual 212 degrees Fahrenheit (100 degrees Celsius). About two-thirds of the operating nuclear reactor power plants in the United States are PWRs.

Probabilistic risk assessment (PRA)

A systematic method for assessing three questions that the NRC uses to define "risk." These questions consider (1) what can go wrong, (2) how likely it is to happen, and (3) what the consequences might be. These questions allow the NRC to understand likely outcomes, sensitivities, areas of importance, system interactions, and areas of uncertainty, which the staff can use to identify risk-significant scenarios. The NRC uses PRA to determine a numeric estimate of risk to provide insights into the strengths and weaknesses of the design and operation of a nuclear power plant.

Production expense

Production expense is one component of the cost of generating electric power, which includes costs associated with fuel, as well as plant operation and maintenance.

Rad (radiation absorbed dose)

One of the two units used to measure the amount of radiation absorbed by an object or person, known as the "absorbed dose," which reflects the amount of energy that radioactive sources deposit in materials through which they pass. The radiation-absorbed dose (rad) is the amount of energy (from any type of ionizing radiation) deposited in any medium (e.g., water, tissue, air). An absorbed dose of 1 rad means that 1 gram of material absorbed 100 ergs of energy (a small but measurable amount) as a result of exposure to radiation. The related international system unit is the gray (Gy), where 1 Gy is equivalent to 100 rad.

Radiation, ionizing

A form of radiation, which includes alpha particles, beta particles, gamma rays and x-rays, neutrons, high-speed electrons, and high-speed protons. Compared to nonionizing radiation, such as found in ultraviolet light or microwaves, ionizing radiation is considerably more energetic. When ionizing radiation passes through material such as air, water, or living tissue, it deposits enough energy to break molecular bonds and displace (or remove) electrons. This electron displacement may lead to changes in living cells. Given this ability, ionizing radiation has a number of beneficial uses, including treating cancer or sterilizing medical equipment. However, ionizing radiation is potentially harmful if not used correctly, and high doses may result in severe skin or tissue damage. It is for this reason that the NRC strictly regulates commercial and institutional uses of the various types of ionizing radiation.

Radiation, nuclear

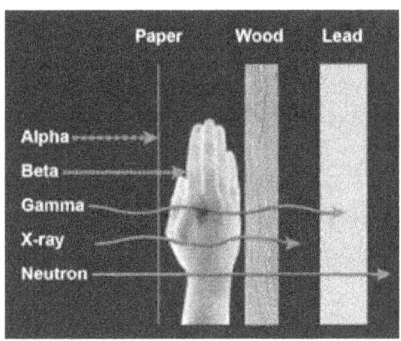

Energy given off by matter in the form of tiny fast-moving particles (alpha particles, beta particles, and neutrons) or pulsating electromagnetic rays or waves (gamma rays) emitted from the nuclei of unstable radioactive atoms. All matter is composed of atoms, which are made up of various parts; the nucleus contains minute particles called protons and neutrons, and the atom's outer shell contains other particles called electrons. The nucleus carries a positive electrical charge, while the electrons carry a negative electrical charge. These forces work toward a strong, stable balance by getting rid of excess atomic energy (radioactivity). In that process, unstable radioactive nuclei may emit energy, and this spontaneous emission is called nuclear radiation. All types of nuclear radiation are also ionizing radiation, but the reverse is not necessarily true; for example, x-rays are a type of ionizing radiation, but they are not nuclear radiation, because they do not originate from atomic nuclei. In addition, some elements are naturally radioactive, as their nuclei emit nuclear radiation as a result of radioactive decay, but others become radioactive by being irradiated in a reactor. Naturally occurring nuclear radiation is indistinguishable from induced radiation.

Radiation source

A radioactive material or byproduct that is specifically manufactured or obtained for the purpose of using the emitted radiation. Such sources are commonly used in teletherapy or industrial radiography; in various types of industrial gauges, irradiators, and gamma knives; and as power sources for batteries (such as those used in spacecraft). These sources usually consist of a known quantity of radioactive material, which is encased in a manmade capsule, sealed between layers of nonradioactive material, or firmly bonded to a nonradioactive substrate to prevent radiation leakage. Other radiation sources include devices such as accelerators and x-ray generators.

Radiation standards

Exposure limits; permissible concentrations; rules for safe handling; and regulations regarding receipt, possession, use, transportation, storage, disposal, and industrial control of radioactive material.

Radiation therapy (radiotherapy)

The therapeutic use of ionizing radiation to treat disease in patients. Although most radiotherapy procedures are intended to kill cancerous tissue or reduce the size of a tumor, therapeutic doses may also be used to reduce pain or treat benign conditions. For example, intervascular brachytherapy uses radiation to treat clogged blood vessels. Other common radiotherapy procedures include gamma stereotactic radiosurgery (gamma knife), teletherapy, and iodine treatment to correct an overactive thyroid gland. These procedures use radiation sources, regulated by the NRC and its Agreement States, that may be applied either inside or outside the body. In either case, the goal of radiotherapy is to deliver the required therapeutic or pain-relieving dose of radiation with high precision and for the required length of time, while preserving the surrounding healthy tissue.

Radiation warning symbol

An officially prescribed magenta or black trefoil on a yellow background, which must be displayed where certain quantities of radioactive materials are present or where certain doses of radiation could be received.

Radioactive contamination

Undesirable radioactive material (with a potentially harmful effect) that is either airborne or deposited in (or on the surface of) structures, objects, soil, water, or living organisms (people, animals, or plants) in a concentration that may harm people, equipment, or the environment.

Radioactive decay

The spontaneous transformation of one radioisotope into one or more different isotopes (known as "decay products" or "daughter products"), accompanied by a decrease in radioactivity (compared to the parent material). This transformation takes place over a defined period of time (known as a "half-life"), as a result of electron capture; fission; or the emission of alpha particles, beta particles, or photons (gamma radiation or x-rays) from the nucleus of an unstable atom. Each isotope in the sequence (known as a "decay chain") decays to the next until it forms a stable, less energetic end product. In addition, radioactive decay may refer to gamma-ray and conversion electron emission, which only reduces the excitation energy of the nucleus.

Radioactivity

The property possessed by some elements (such as uranium) of spontaneously emitting energy in the form of radiation as a result of the decay (or disintegration) of an unstable atom. Radioactivity is also the term used to describe the rate at which radioactive material emits radiation. Radioactivity is measured in units of becquerels or disintegrations per second.

Radiography

The use of sealed sources of ionizing radiation for nondestructive examination of the structure of materials. When the radiation penetrates the material, it produces a shadow image by blackening a sheet of photographic film that has been placed behind the material, and the differences in blackening suggest flaws and unevenness in the material.

Radioisotope (radionuclide)

An unstable isotope of an element that decays or disintegrates spontaneously, thereby emitting radiation. Approximately 5,000 natural and artificial radioisotopes have been identified.

Radiopharmaceutical

A pharmaceutical drug that emits radiation and is used in diagnostic or therapeutic medical procedures. Radioisotopes that have short half-lives are generally preferred to minimize the radiation dose to the patient and the risk of prolonged exposure. In most cases, these short-lived radioisotopes decay to stable elements within minutes, hours, or days, allowing patients to be released from the hospital in a relatively short time.

Reactor core

The central portion of a nuclear reactor, which contains the fuel assemblies, water, and control mechanisms, as well as the supporting structure. The reactor core is where fission takes place.

Reactor Oversight Process (ROP)

The process by which the NRC monitors and evaluates the performance of commercial nuclear power plants. Designed to focus on those plant activities that are most important to safety, the ROP uses inspection findings and performance indicators to assess each plant's safety performance.

Regulation

The governmental function of controlling or directing economic entities through the process of rulemaking and adjudication.

Regulatory Information Conference

An annual NRC conference that brings together NRC staff, regulated utilities, materials users, and other interested stakeholders to discuss nuclear safety topics and significant and timely regulatory activities through informal dialogue to ensure an open regulatory process.

Rem (roentgen equivalent man)

One of the two standard units used to measure the dose equivalent (or effective dose), which combines the amount of energy (from any type of ionizing radiation) that is deposited in human tissue with the biological effects of the given type of radiation. For beta and gamma radiation, the dose equivalent is the same as the absorbed dose. By contrast, the dose equivalent is larger than the absorbed dose for alpha and neutron radiation, because these types of radiation are more damaging to the human body. Thus, the dose equivalent (in rems) is equal to the absorbed dose (in rads) multiplied by the quality factor of the type of radiation (Title 10 of the *Code of Federal Regulations* (10 CFR) 20.1004, "Units of Radiation Dose"). The related international system unit is the sievert (Sv), where 100 rem is equivalent to 1 Sv.

Renewable resources

Natural, but limited, energy resources that can be replenished, including biomass, hydro, geothermal, solar, and wind. These resources are virtually inexhaustible but limited in the amount of energy that is available per unit of time. In the future, renewable resources could also include the use of ocean thermal, wave, and tidal action technologies. Utility renewable resource applications include bulk electricity generation, onsite electricity generation, distributed electricity generation, nongrid-connected generation, and demand-reduction (energy efficiency) technologies.

Risk

The combined answer to three questions that consider (1) what can go wrong, (2) how likely it is to occur, and (3) what the consequences might be. These three questions allow the NRC to understand likely outcomes, sensitivities, areas of importance, system interactions, and areas of uncertainty, which can be used to identify risk-significant scenarios.

Risk-based decisionmaking

An approach to regulatory decisionmaking that considers only the results of a probabilistic risk assessment.

Risk-informed decisionmaking

An approach to regulatory decisionmaking, in which insights from probabilistic risk assessment are considered with other engineering insights.

Risk-informed regulation

An approach to regulation taken by the NRC, which incorporates an assessment of safety significance or relative risk. This approach ensures that the regulatory burden imposed by an individual regulation or process is appropriate to its importance in protecting the health and safety of the public and the environment.

Risk significant

"Risk significant" can refer to a facility's system, structure, component, or accident sequence that exceeds a predetermined limit for contributing to the risk associated with the facility. The term also describes a level of risk exceeding a predetermined "significance" level.

Safeguards

The use of material control and accounting programs to verify that all special nuclear material is properly controlled and accounted for, as well as the physical protection (or physical security) equipment and security forces. As used by IAEA, this term also means verifying that the peaceful use commitments made in binding nonproliferation agreements, both bilateral and multilateral, are honored.

Safeguards Information

A special category of sensitive unclassified information that must be protected. Safeguards Information concerns the physical protection of operating power reactors, spent fuel shipments, strategic special nuclear material, or other radioactive material.

Safety related

In the regulatory arena, this term applies to systems, structures, components, procedures, and controls (of a facility or process) that are relied upon to remain functional during and following design-basis events. Their functionality ensures that key regulatory criteria, such as levels of radioactivity released, are met. Examples of safety-related functions include shutting down a nuclear reactor and maintaining it in a safe-shutdown condition.

Safety significant

When used to qualify an object, such as a system, structure, component, or accident sequence, this term identifies that object as having an impact on safety, whether determined through risk analysis or other means, that exceeds a predetermined significance criterion.

SAFSTOR

A method of decommissioning in which a nuclear facility is placed and maintained in a condition that allows the facility to be safely stored and subsequently decontaminated (deferred decontamination) to levels that permit release for unrestricted use.

Scram

The sudden shutting down of a nuclear reactor, usually by rapid insertion of control rods, either automatically or manually by the reactor operator (also known as a "reactor trip").

Sensitive unclassified nonsafeguards information

Information that is generally not publicly available and that encompasses a wide variety of categories, such as proprietary information, personal and private information, or information subject to attorney-client privilege.

Shutdown

A decrease in the rate of fission (and heat or energy production) in a reactor (usually by the insertion of control rods into the core).

Source material

Uranium or thorium, or any combination thereof, in any physical or chemical form, or ores that contain, by weight, 1/20 of 1 percent (0.05 percent) or more of (1) uranium, (2) thorium, or (3) any combination thereof. Source material does not include special nuclear material.

Special nuclear material

Plutonium, uranium-233, or uranium enriched in the isotopes uranium-233 or uranium-235.

Spent fuel pool

An underwater storage and cooling facility for spent (depleted) fuel assemblies that have been removed from a reactor.

Spent (depleted or used) nuclear fuel

Nuclear reactor fuel that has been used to the extent that it can no longer effectively sustain a chain reaction.

Subcriticality

The condition of a nuclear reactor system, in which nuclear fuel no longer sustains a fission chain reaction (that is, the reaction fails to initiate its own repetition, as it would in a reactor's normal operating condition). A reactor becomes subcritical when its fission events fail to release a sufficient number of neutrons to sustain an ongoing series of reactions, possibly as a result of increased neutron leakage or poisons.

Teletherapy

Treatment in which the source of the therapeutic radiation is at a distance from the body. Because teletherapy is often used to treat malignant tumors deep within the body by bombarding them with a high-energy beam of gamma rays (from a radioisotope such as cobalt-60) projected from outside the body, it is often called "external beam radiotherapy."

Title 10 of the *Code of Federal Regulations* (10 CFR)

Four volumes of the *Code of Federal Regulations* (CFR) address energy-related topics. Parts 1 to 199 contain the regulations (or rules) established by the NRC. These regulations govern the transportation and storage of nuclear materials; use of radioactive materials at nuclear power plants, research and test reactors, uranium recovery facilities, fuel cycle facilities, waste repositories, and other nuclear facilities; and use of nuclear materials for medical, industrial, and academic purposes.

Transient

A change in the reactor coolant system temperature, pressure, or both, attributed to a change in the reactor's power output. Transients can be caused by (1) adding or removing neutron poisons, (2) increasing or decreasing electrical load on the turbine generator, or (3) accident conditions.

Transuranic waste

Material contaminated with transuranic elements—artificially made, radioactive elements, such as neptunium, plutonium, americium, and others—that have atomic numbers higher than uranium in the periodic table of elements. Transuranic waste is primarily produced from recycling spent fuel or using plutonium to fabricate nuclear weapons.

Tritium

A radioactive isotope of hydrogen. Because it is chemically identical to natural hydrogen, tritium can easily be taken into the body by any ingestion path. It decays by emitting beta particles and has a half-life of about 12.5 years.

Uprate

See *Power uprate*.

Uranium

A radioactive element with the atomic number 92 and, as found in natural ores, an atomic weight of approximately 238. The two principal natural isotopes are uranium-235 (which comprises 0.7 percent of natural uranium), which is fissile, and uranium-238 (99.3 percent of natural uranium), which is fissionable by fast neutrons and is fertile, meaning that it becomes fissile after absorbing one neutron. Natural uranium also includes a minute amount of uranium-234.

Uranium enrichment

The process of increasing the percentage of U235 from 0.7 percent in natural uranium to about 3-5 percent for use in fuel for nuclear reactors. Enrichment can be done through gaseous diffusion, gas centrifuges, or laser isotope separation.

Uranium fuel fabrication facility

A facility that converts enriched UF_6 into fuel for commercial light-water power reactors, research and test reactors, and other nuclear reactors. The UF_6, in solid form in containers, is heated to a gaseous form and then chemically processed to form uranium dioxide (UO_2) powder. This powder is then processed into ceramic pellets and loaded into metal tubes, which are subsequently bundled into fuel assemblies. Fabrication also can involve MOX fuel, which contains plutonium oxide mixed with either natural or depleted uranium oxide, in ceramic pellet form.

Uranium hexafluoride production facility (or uranium conversion facility)

A facility that receives natural uranium in the form of ore concentrate (known as yellowcake) and converts it into UF_6, in preparation for fabricating fuel for nuclear reactors.

U.S. Department of Energy (DOE)

The Federal agency established by Congress to advance the national, economic, and energy security of the United States, among other missions.

U.S. Department of Homeland Security (DHS)

The Federal agency responsible for leading the unified national effort to secure the United States against those who seek to disrupt the American way of life. DHS is also responsible for preparing for and responding to all hazards and disasters and includes the formerly separate FEMA, the Coast Guard, and the Secret Service.

U.S. Environmental Protection Agency (EPA)

The Federal agency responsible for protecting human health and safeguarding the environment. The EPA leads the Nation's environmental science, research, education, and assessment efforts to ensure that attempts to reduce environmental risk are based on the best available scientific information. The EPA also ensures that environmental protection is an integral consideration in U.S. policies.

Viability assessment

A decisionmaking process used by DOE to assess the prospects for safe and secure permanent disposal of HLW in an excavated, underground facility, known as a geologic repository. This decisionmaking process is based on (1) specific design work on the critical elements of the repository and waste package, (2) a total system performance assessment that will describe the probable behavior of the repository, (3) a plan and cost estimate for the work required to complete the license application, and (4) an estimate of the costs to construct and operate the repository.

Glossary

Waste, radioactive

Radioactive materials at the end of their useful life or in a product that is no longer useful and requires proper disposal.

Waste classification (classes of waste)

Classification of LLW according to its radiological hazard. The classes include Class A, B, and C, with Class A being the least hazardous and accounting for 96 percent of LLW. As the waste class and hazard increase, the regulations established by the NRC require progressively greater controls to protect the health and safety of the public and the environment.

Watt

A unit of power (in the international system of units) defined as the consumption or conversion of 1 joule of energy per second. In electricity, a watt is equal to current (in amperes) multiplied by voltage (in volts).

Watthour

An unit of energy equal to 1 watt of power steadily supplied to, or taken from, an electrical circuit for 1 hour (or exactly 3.6×10^3 joules).

Well logging

All operations involving the lowering and raising of measuring devices or tools that contain licensed nuclear material or are used to detect licensed nuclear materials in wells for the purpose of obtaining information about the well or adjacent formations that may be used in oil, gas, mineral, ground water, or geological exploration.

Wheeling service

The movement of electricity from one system to another over transmission facilities of intervening systems. Wheeling service contracts can be established between two or more systems.

Yellowcake

The solid form of mixed uranium oxide, which is produced from uranium ore in the uranium recovery (milling) process. The material is a mixture of uranium oxides, which can vary in proportion and color from yellow to orange to dark green (blackish) depending on the temperature at which the material is dried (which affects the level of hydration and impurities), with higher drying temperatures producing a darker and less soluble material. (The yellowcake produced by most modern mills is actually brown or black, rather than yellow, but the name comes from the color and texture of the concentrates produced by early milling operations.) Yellowcake is commonly referred to as U_3O_8, because that chemical compound comprises approximately 85 percent of the yellowcake produced by uranium recovery facilities, and that product is then transported to a uranium conversion facility, where it is transformed into UF_6, in preparation for fabricating fuel for nuclear reactors.

Zirconium

A chemical element used (in the form of "zircaloy" metals) in cladding for nuclear fuel rods. The thin zirconium tubes contain pellets of nuclear fuel and are bundled together into assemblies for use in a reactor.

Web Link Index

Worldwide Electricity Generated by Commercial Nuclear Power

International Atomic Energy Agency (IAEA)
www.iaea.org

IAEA Power Reactor Information System (PRIS)
www.iaea.org/programmes/a2

Nuclear Energy Agency (NEA)
www.nea.fr/

World Nuclear Association (WNA)
www.world-nuclear.org/

World Nuclear Power Reactors and Uranium Requirements
www.world-nuclear.org/info/reactors.html

WNA Reactor Database
www.world-nuclear.org/reference/default.aspx

WNA Global Nuclear Reactors Map
www.wano.org.uk/WANO_Documents/WANO_Map/WANO_Map.pdf

NRC Office of International Programs
www.nrc.gov/about-nrc/organization/oipfuncdesc.html

NRC Regulatory Information Conference (RIC)
www.nrc.gov/public-involve/conference-symposia/ric/index.html

International Activities

Treaties and Conventions
www.nrc.gov/about-nrc/ip/treaties-conventions.html

Operating Nuclear Reactors

U.S. Commercial Nuclear Power Reactors

Commercial Reactors
www.nrc.gov/info-finder/reactor/

Oversight of U.S. Commercial Nuclear Power Reactors

Reactor Oversight Process (ROP)
www.nrc.gov/NRR/OVERSIGHT/ASSESS/index.html

NUREG-1649, "Reactor Oversight Process"
www.nrc.gov/reading-rm/doc-collections/nuregs/staff/sr1649

ROP Performance Indicators Summary
www.nrc.gov/NRR/OVERSIGHT/ASSESS/pi_summary.html

Future U.S. Commercial Nuclear Power Reactor Licensing

New Reactor License Process
www.nrc.gov/reactors/new-reactor-op-lic/licensing-process.html#licensing

New Reactors

New Reactor Licensing
www.nrc.gov/reactors/new-reactors.html

Reactor License Renewal

Reactor License Renewal Process
www.nrc.gov/reactors/operating/licensing/renewal/process.html

10 CFR Part 51, "Environmental Protection Regulations for Domestic Licensing and Related Regulatory Functions"
www.nrc.gov/reading-rm/doc-collections/cfr/part051/

10 CFR Part 54, "Requirements for Renewal of Operating Licenses for Nuclear Power Plants"
www.nrc.gov/reading-rm/doc-collections/cfr/part054/

Status of License Renewal Applications and Industry Activities
www.nrc.gov/reactors/operating/licensing/renewal/applications.html

U.S. Nuclear Research and Test Reactors

Research and Test Reactors
www.nrc.gov/reactors/non-power.html

Nuclear Regulatory Research

Nuclear Reactor Safety Research
www.nrc.gov/about-nrc/regulatory/research/reactor-rsch.html

State-of-the-Art Reactor Consequence Analyses (SOARCA)
www.nrc.gov/about-nrc/regulatory/research/soar.html

Risk Assessment in Regulation
www.nrc.gov/about-nrc/regulatory/risk-informed.html

Digital Instrumentation and Controls
www.nrc.gov/about-nrc/regulatory/research/digital.html

Computer Codes
www.nrc.gov/about-nrc/regulatory/research/comp-codes.html

Generic Issues Program
www.nrc.gov/about-nrc/regulatory/gen-issues.html

The Committee To Review Generic Requirements (CRGR)
www.nrc.gov/about-nrc/regulatory/crgr.html

Nuclear Materials

U.S. Fuel Cycle Facilities

U.S. Fuel Cycle Facilities
www.nrc.gov/info-finder/materials/fuel-cycle/

Uranium Recovery

Uranium Milling/Recovery
www.nrc.gov/info-finder/materials/uranium/

U.S. Materials Licenses

Materials Licensees Toolkits
www.nrc.gov/materials/miau/mat-toolkits.html

Medical Applications

Medical Applications
www.nrc.gov/materials/medical.html

Medical Uses

Medical Uses
www.nrc.gov/materials/miau/med-use.html

Nuclear Gauges and Commercial Product Irradiators

General Licenses Uses
www.nrc.gov/materials/miau/general-use.html

Industrial Uses of Nuclear Material

Industrial Applications
www.nrc.gov/materials/miau/industrial.html

Exempt Consumer Products
www.nrc.gov/materials/miau/consumer-pdts.html

Radioactive Waste

U.S. Low-Level Radioactive Waste Disposal

Low-Level Radioactive Waste
www.nrc.gov/waste/low-level-waste.html

U.S. High-Level Radioactive Waste Management: Disposal and Storage

High-Level Radioactive Waste
www.nrc.gov/waste/high-level-waste.html

Spent Nuclear Fuel Storage

Spent Nuclear Fuel Storage
www.nrc.gov/waste/spent-fuel-storage.html

U.S. Nuclear Materials Transportation

Nuclear Materials Transportation
www.nrc.gov/materials/transportation.html

Decommissioning

Decommissioning
www.nrc.gov/about-nrc/regulatory/decommissioning.html

Nuclear Security and Emergency Preparedness

Nuclear Security
www.nrc.gov/security.html

Domestic Safeguards

Domestic Safeguards
www.nrc.gov/security/domestic.html

Information Security

Information Security
www.nrc.gov/security/info-security.html

Ensuring the Security of Radioactive Material
www.nrc.gov/security/byproduct.html

Emergency Preparedness and Response

Emergency Preparedness and Response
www.nrc.gov/about-nrc/emerg-preparedness.html

Research and Test Reactor Emergency Preparedness

Research and Test Reactors
www.nrc.gov/reactors/non-power.html

Emergency Preparedness Stakeholder Meetings and Workshops

www.nrc.gov/public-involve/public-meetings/stakeholder-mtngs-wksps.html

Emergency Action Level Development

www.nrc.gov/about-nrc/emerg-preparedness/about-emerg-preparedness/emerg-action-level-dev.html

Hostile-Action-Based Emergency Preparedness Drills

www.nrc.gov/about-nrc/emerg-preparedness/respond-to-emerg/hostile-action.html

Exercise Schedules

NRC Participation Exercise Schedule
www.nrc.gov/about-nrc/emerg-preparedness/about-emerg-preparedness/exercise-schedules.html

Other Web Links

Employment Opportunities

NRC—*A Great Place to Work*
www.nrc.gov/about-nrc/employment.html

Glossary

NRC Basic References
www.nrc.gov/reading-rm/basic-ref/glossary/full-text.html

Glossary of Electricity Terms

www.eia.doe.gov/cneaf/electricity/epav1/glossary.html

Glossary of Security Terms

https://hseep.dhs.gov/DHSResource/Glossary.aspx

Public Involvement

NRC Library
www.nrc.gov/reading-rm.html

Freedom of Information and Privacy Acts
www.nrc.gov/reading-rm/foia/foia-privacy.html

Agencywide Documents Access and Management System (ADAMS)
www.nrc.gov/reading-rm/adams.html

Public Document Room
www.nrc.gov/reading-rm/pdr.html

Public Meeting Schedule
www.nrc.gov/public-involve/public-meetings/index.cfm

Documents for Comment
www.nrc.gov/public-involve/doc-comment.html

Small Business and Civil Rights

Contracting Opportunities for Small Businesses
www.nrc.gov/about-nrc/contracting/small-business.html

Workplace Diversity
www.nrc.gov/about-nrc/employment/diversity.html

Discrimination Complaint Activity
www.nrc.gov/about-nrc/civil-rights.html

Equal Employment Opportunity Program
www.nrc.gov/about-nrc/civil-rights/eeo.html

Limited English Proficiency
www.nrc.gov/about-nrc/civil-rights/limited-english.html

Minority Serving Institutions Program
www.nrc.gov/about-nrc/grants.html#msip

NRC Comprehensive Diversity Management Plan brochure
www.nrc.gov/reading-rm/doc-collections/nuregs/brochures/br0316

Index

NRC FORM 335 (2-89) NRCM 1102, 3201,3202	U.S. NUCLEAR REGULATORY COMMISSION BIBLIOGRAPHIC DATA SHEET *(See instructions on the reverse)*	1. REPORT NUMBER (Assigned by NRC, Add Vol., Supp., Rev., and Addendum Numbers, If any.) NUREG-1350, Vol. 24

2. TITLE AND SUBTITLE	3. DATE REPORT PUBLISHED	
U.S. Nuclear Regulatory Commission Information Digest 2012-2013	MONTH	YEAR
	August	2012
	4. FIN OR GRANT NUMBER	
	n/a	

5. AUTHOR(S)	6. TYPE OF REPORT
Ivonne Couret, et al.	Annual
	7. PERIOD COVERED *(Inclusive Dates)* 2012-2013

8. PERFORMING ORGANIZATION - NAME AND ADDRESS *(If NRC, provide Division, Office or Region, U.S. Nuclear Regulatory Commission, and mailing address; if contractor, provide name and mailing address.)*

Public Affairs Staff
Office of Public Affairs
U.S. Nuclear Regulatory Commission
Washington, DC 20555-0001

9. SPONSORING ORGANIZATION - NAME AND ADDRESS *(If NRC, type "Same as above"; if contractor, provide NRC Division, Office or Region, U.S. Nuclear Regulatory Commission, and mailing address.)*

Same as 8, above

10. SUPPLEMENTARY NOTES
There may be a supplementary document produced with only the updated figures, tables, and/or appendices within the next 12 months.

11. ABSTRACT *(200 words or less)*

The U.S. Nuclear Regulatory Commission (NRC) 2012–2013 Information Digest provides a summary of information about the NRC and the industry it regulates. It describes the agency's regulatory responsibilities and licensing activities and also provides general information on nuclear-related topics. It is updated annually.

The Information Digest includes NRC and industry data in a quick reference format for activities through 2011 or most recent current data available at manuscript completion. InfoGraphics have been incorporated to help provide a visual representation of information expanding the agency's efforts to be more transparent and broaden its public outreach. The Web Link Index provides URL addresses for more information on major topics. The Digest also includes a tear-out reference sheet, the NRC Facts at a Glance.

The NRC reviewed information from industry and international sources but did not perform an independent verification. This edition contains adjustments to preliminary figures from the previous year. All information is final unless otherwise noted. The NRC is the source for all photographs, graphics, and tables unless otherwise noted.

The agency welcomes comments or suggestions on the Information Digest. Please contact the Office of Public Affairs, U.S. Nuclear Regulatory Commission, Washington, DC 20555-0001 or by e-mail at OPA.Resource@nrc.gov.

12. KEY WORDS/DESCRIPTORS *(List words or phrases that will assist researchers in locating the report.)*	13. AVAILABILITY STATEMENT
Information Digest 2012-2013 Edition NRC Facts Nuclear Regulatory Commission	unlimited
	14. SECURITY CLASSIFICATION
	(This Page) unclassified
	(This Report) unclassified
	15. NUMBER OF PAGES 216
	16. PRICE

NRC FORM 335 (9-2004)

PRINTED ON RECYCLED PAPER

U.S. Nuclear Regulatory Commission
NUREG-1350, Volume 24
August 2012